W9-AXM-765

Calculator Logic Systems and Mathematical Understandings

Calculator Logic Systems and Mathematical Understandings

Enid R. Burrows
Plymouth State College
Plymouth, New Hampshire

National Council of Teachers of Mathematics

Library of Congress Cataloging-in-Publication Data:

Burrows, Enid R.
Calculator logic systems and mathematical understandings / by
Enid R. Burrows.
p. cm.
ISBN 0-87353-295-3
1. Programmable calculators. I. Title.
QA75.B84 1990 90-6090
510′.28—dc20 CIP

The publications of the National Council of Teachers of Mathematics present a variety of viewpoints. The views expressed or implied in this publication, unless otherwise noted, should not be interpreted as official positions of the Council.

Printed in the United States of America

Contents

Preface

In the last eight to ten years, in-service teachers have attended my workshops and lectures on calculator logic systems and calculator algorithms. Concurrently, the mathematics department at Plymouth State College began offering a one-credit module entitled Introduction to Calculators and Computers. This minicourse is required of certain students in mathematics education and is an elective for students in business and liberal arts. *Calculator Logic Systems and Mathematical Understandings* is an outgrowth of these contacts with in-service and preservice teachers of mathematics and is in response to their requests for a book on the subject.

A study of calculator logic systems provides a review and clarification of binary operations and functions, the two pivotal concepts of arithmetic and algebra. For example, since the symbol "−" is customarily used for both the binary operation of subtraction and for the function of the additive inverse, the two concepts are often confused. On the calculator the key for the binary operation of subtraction is [−]; the key for the additive inverse (change sign) function is [+/−] or [CHS].

Two logic systems, algebraic logic and algebraic operating system, are contrasted in chapters 1 and 4. The differences in the treatment of the order of operation of binary operations between these two types of calculators forces one to examine varying strategies or algorithms for the calculation of the same mathematical expression.

In chapter 2 the built-in functions of the calculator are discussed. Many of us have drawn pictures of a function as a "black box" operator on an element from the domain to yield a functional value:

$$x \rightarrow \boxed{f} \rightarrow f(x)$$

The calculator is a prototype of this concept of a function. The early pages of chapter 2 limit the functions discussed to functions that are familiar to students in middle school. Trigonometric and logarithmic functions are discussed at the end of chapter 2. The [INV] key on the calculator reinforces the concept of inverse function. Through further experimentation with the

INV key we can examine the necessity to limit the domain of functions so that they are one to one in order to have the inverse well defined.

Through much of chapters 1–4, functions are contrasted with binary operations. In chapter 5, a discussion of Reverse Polish Notation (RPN) calculator logic allows us to show how a binary operation can be described as a function of two variables.

Throughout the monograph, calculator algorithms are discussed for convenient calculation of typical mathematical expressions. From my experience teaching algebra and statistics courses to college students, I know that although many students can solve a problem correctly and set up the correct expression, they frequently cannot calculate the correct decimal answer with their calculator. Chapter 7 shows a convenient technique for evaluating polynomials with a calculator. This concluding chapter shows the solution of problems that one would not wish to attempt without the aid of some calculating device.

There are many people to thank for their contributions to this project. At Plymouth State College, students helped to shape my early thinking and used preliminary rough drafts, whereas colleagues who used later revisions made valuable suggestions. The NCTM editorial staff has coordinated reviews of the manuscript and made useful recommendations. The final product has benefited from this input. Finally, my daughters, Robin and Lanette, have encouraged and supported me throughout this project. Thank you!

Introduction

THIS monograph is aimed at helping the reader to understand the built-in logic of various calculator operating systems. The mathematical concepts of function and binary operation are constantly used and reviewed to help readers design effective calculator algorithms for evaluating a variety of mathematical expressions. This review of mathematical concepts couched in the nontraditional setting of the calculator is an equally important goal.

The topics addressed reflect many of the issues raised in the NCTM's *Curriculum and Evaluation Standards for School Mathematics,* hereafter referred to as the *Standards*. Although the *Standards* assumes that calculators will be available to all students at all times, it notes that "unfortunately, most students are not educated to participate in this new [technological] society" (p. 4). The same can also be said of many preservice and in-service mathematics teachers.

Two themes are particularly stressed. The first is the need to teach the general public, as well as mathematics professionals, that not all calculators operate in the same way. Indeed, exactly the same keystroke sequence performed on different calculators may result in different answers in the display. This variation in operating systems has some pedagogical advantages in that it provides an opportunity to discover patterns and to discuss mathematical concepts. As the *Standards* puts it, "pattern recognition is at the heart of mathematical thinking" (p. 9).

The second theme is the need to teach how to construct appropriate algorithms for calculator computation. My experience with college students indicates that although the mathematical expression they use for the solution of a problem may be correct, they may not evaluate it correctly using a calculator. Many times they use algorithms that may be more appropriate for paper-and-pencil calculation than for calculator evaluation.

Most examples in this monograph are addressed to the pivotal problem-solving issue of the kinds of algorithms that can be designed for calculator computation of various mathematical expressions. The ability to design appropriate algorithms, of course, assumes an understanding of the first point that calculators may differ with respect to operating systems.

The *Standards* describes problem-solving situations as those in which "it is assumed that students have acquired the necessary concepts and procedures to find a solution but need to use strategies (heuristics) to make the connections between the given information and a method of solution" (p. 9). In this monograph the *solution* to a problem is defined as an algorithm or procedure and not a numerical answer. Thus it may not be unique, since there may be several possible strategies for calculating a mathematical expression. An algorithm is a sequence of keystrokes comprising a strategy for arriving at a numerical answer. In discussing such strategies, we must assume that mathematical concepts such as function, inverse, and binary operation are familiar to the reader.

In stressing the points that a strategy appropriate for one operating system may be inappropriate for others and that there may be several useful strategies to choose from within a given operating system, we are reinforcing the position in the *Standards* (in the K–4 guidelines, but appropriate for all levels) that "students should be led to see that a variety of processes can be used to determine the answer" (p. 23). There are frequent occasions in this monograph where alternative strategies are contrasted and compared both within and between operating systems.

Formulating calculator algorithms engages students in all the processes that the *Standards* describes as involved in problem solving: "making conjectures, investigating and exploring ideas, discussing and questioning their own thinking and the thinking of others, validating results, and making convincing arguments" (p. 54).

1
Binary Operations in Algebraic Logic (AL) and Algebraic Operating Systems (AOS)

SUPPOSE one attempts to push the following calculator keys:

8
[−]
3
[×]
4
[=]

It is possible to find four different types of calculators. One calculator may have no [=] key. The next calculator may have two or even three [=] keys. One calculator will show -4 in the display, whereas another will show 20. Unraveling a rationale for these divergent display results and exploring ways in which different calculators may be effectively used is what this book is about.

We will discuss three major operating systems: Algebraic Logic (AL), Algebraic Operating System (AOS)[1], and Reverse Polish Notation (RPN). A fourth type of calculator using *arithmetic logic* is really a combination of an algebraic logic calculator and a Reverse Polish Notation calculator. Since the study of these different types of calculators involves understanding how numbers are combined through binary operations (functions of two variables) and how numbers are acted upon by a function of a single variable (unary operations), we will assume and sometimes review a basic knowledge of algebra such as might be learned from a two-year high school course.

1. Algebraic Operating System and AOS are registered trademarks of Texas Instruments. Some Texas Instruments models operate with AL; other models operate with AOS. Other manufacturers also have calculators with these varying operating characteristics.

It will be important to distinguish between a mathematical expression and a keystroke sequence that is used to evaluate that expression on any given calculator. We will also want to distinguish between the value of the mathematical expression and the numbers that are displayed on the calculator.

The keystroke sequence is the sequence of keys to be pushed on a given calculator in computing a mathematical expression. It will be denoted by a vertical representation, and the mathematical expression will be shown by the usual horizontal representation. See example 1.

Example 1

Mathematical expression	*Keystroke sequence for AL or AOS*
$\dfrac{3 \times 5 \times 9}{7 \times 4 \times 6}$	3 [×] 5 [×] 9 [÷] 7 [÷] 4 [÷] 6 [=]

It should be noted that this keystroke sequence calls for dividing by 4 and dividing by 6, not multiplying. One can verify that this is correct by rewriting the mathematical expression as:

$$(3) \times (5) \times (9) \times (1/7) \times (1/4) \times (1/6) =$$

and noting that multiplication by the reciprocal of a number is equivalent to division by the number.

Consider the keystroke sequence shown at the right. The sequence of pushing these five keys enters the number 456.7 on all calculators. A keystroke sequence of successive numbers and decimals will be represented by the number it enters in the display. See example 2.

4
6
5
.
7

Of course the keystroke sequence indicated in example 2 requires the pushing of eight (not four) keys to complete.

The keystroke sequence needed to evaluate a given mathematical expression depends on the type of calculator used. The examples given above are keystroke sequences that work on algebraic logic calculators or algebraic

Example 2

Mathematical expression	*Keystroke sequence in AL or AOS*
46.2 × 30	46.2 [×] 30 [=]

operating system calculators, both of which have an [=] key. They could not be used with the RPN calculator, which does not have an [=] key. Later we will present examples of keystroke sequences that give entirely different display results for AL and AOS calculators.

The three types of operating systems differ primarily in their treatment of binary operations. The four well-known binary operations are addition (+), subtraction (−), multiplication (×), and division (÷). In addition to these four operations some calculators also include a key for exponentiation (y^x). A binary operation combines two numbers into a single number. For example, in 4 + 3 = 7 the numbers 4 and 3 are combined by the binary operation of addition into the single number 7. For both algebraic logic (AL) calculators and algebraic operating system (AOS) calculators, a single binary operation is performed by entering the keystroke sequence in the same order as it would be written in its mathematical expression. For example:

Mathematical expression	8 − 5 =	34 × 6 =	98 / 14 =	2^3 =
Keystroke sequence in AL or AOS	8 [−] 5 [=]	34 [×] 6 [=]	98 [÷] 14 [=]	2 [y^x] 3 [=]
Answer/display	3	204	7	8

The AL and AOS calculators differ in their processing when binary operations are performed one after another. *The AL calculator processes the operations in the order in which they are entered in the keystroke sequence.* Thus for the keystroke sequence listed below at the right, the AL calculator would evaluate the sequence as follows:

2 + 4 × 5 =
6 × 5 = 30 (shown in display on AL)

2 [+] 4 [×] 5 [=]

This would be the keystroke sequence used in AL to calculate the mathematical expression $(2 + 4) \times 5$. *The AOS calculator processes the operations according to algebraic hierarchy by which operations are performed in the following order:* (1) evaluation of parentheses (not actually a binary operation), (2) exponentiation, (3) multiplication and division left to right (or top to bottom), and (4) addition and subtraction left to right (or top to bottom). For the keystroke sequence shown at the right above, the AOS calculator would evaluate the keystroke sequence as follows:

$2 + 4 \times 5 =$
$2 + 20 = 22$ (shown in display on AOS)

This keystroke sequence on an AOS calculator computes the mathematical expression $2 + 4 \times 5$. Thus we see that entering exactly the same keystroke sequence on two different calculators may result in completely different display results. *The easiest way to determine whether your calculator, which has an* [=] *key, is an AL calculator or an AOS calculator is to enter a keystroke sequence similar to the one above in which addition (or subtraction) is keyed prior to multiplication (or division) and to infer from the result displayed which operation is being performed first.*

It should be noted that multiplication and division are performed top to bottom (or left to right) in either AOS or AL—not multiplication and then division as is often assumed. Hence,

$$12 \div 3 \times 2 = 8, \textit{ not } 2.$$

This calculator operation is consistent with standard algebraic hierarchy and expression, but is often confused with the expression

$$\frac{12}{3 \times 2} = 12 \div (3 \times 2) = 12 \div 3 \div 2.$$

Example 3 shows how keystrokes can differ for the same calculation using AL as opposed to AOS:

Example 3

Mathematical expression	*AL*	*AOS*
$(3 + 4) \times 5$	3 [+] 4 [×] 5 [=]	3 [+] 4 [=] [×] 5 [=]

Notice that the [=] key is used in the AOS keystroke sequence above to complete the pending operation of addition before the operation of multiplication is performed. One could have also used the parenthesis keys, but that would have added another keystroke to the sequence needed to complete the computation. A second, probably more compelling reason not to use the parenthesis keys is that the number of pending operations that any AOS calculator can retain is finite and so it is advisable to avoid the unnecessary use of parenthesis. Examining some additional examples will illustrate the differences between AL and AOS and will develop strategies that might be used with each operating system.

Example 4

Mathematical expression	*AL*	*AOS*
$3 + 4 \times 5$	4 [×] 5 [+] 3 [=]	3 [+] 4 [×] 5 [=]

Of course, to compute this expression the AL sequence could be used with the AOS calculator, but not vice versa.

Example 5

Mathematical expression	*AL*	*AOS*
$3 \times 4 - 5 \times 6$	3 [×] 4 [÷] 6 [−] 5 [×] 6 [=]	3 [×] 4 [−] 5 [×] 6 [=]

The AOS keystroke sequence above is straightforward, since operations are performed according to the algebraic hierarchy. The AL keystroke sequence can best be understood if the expression is rewritten as

$$3 \times 4 - 5 \times 6 = \left(\frac{3 \times 4}{6} - 5\right) \times 6 = (3 \times 4 \div 6 - 5) \times 6.$$

A similar scheme can be used to compute the sum or difference of fractions.

Example 6

Mathematical expression	*AL*	*AOS*
$\frac{3}{7} + \frac{5}{6}$	3 [÷] 7 [×] 6 [+] 5 [÷] 6 [=]	3 [÷] 7 [+] 5 [÷] 6 [=]

Of course the display for both types of calculators will be the finite decimal form, 1.2619048, rather than the fractional form, 1 11/42.

Example 7

Mathematical expression	*AL*	*AOS*
$13 - 2 \times 4$	13 [÷] 4 [−] 2 [×] 4 [=]	13 [−] 2 [×] 4 [=]

Although this expression resembles example 4 ($3 + 4 \times 5$), the AL strategy used in example 4 took advantage of the commutative property of addition ($a + b = b + a$). Since subtraction is not commutative the AL keystroke sequence for example 7 follows the algorithm developed in example 5 and 6 for the sum or difference of two products.

In subsequent chapters we will examine alternative keystroke sequences for the solution of this last type of example that use both the addressable memory and the additive inverse function.

Example 8

Mathematical expression	*AL*	*AOS*
$\dfrac{3 \times 5 - 4}{2 \times 7}$	3 [×] 5 [−] 4 [÷] 2 [÷] 7 [=]	3 [×] 5 [−] 4 [=] [÷] 2 [÷] 7 [=]

The keystroke sequences above simply combine algorithms that we have previously discussed, namely: (1) To divide by a product of numbers, divide successively by each factor; (2) When using AOS, use the [=] key to complete pending operations of a lower hierarchy before continuing with the operations of a higher hierarchy.

Review

1. What are the defining characteristics of the algebraic logic (AL) calculator?
2. What are the defining characteristics of the algebraic operating system (AOS) calculator?
3. What key is missing from an RPN (Reverse Polish Notation) calculator that is always found on AL and AOS calculators?
4. Without referring to the operating manual (which may not tell you anyway), how can you determine what type of operating system a calculator uses?

Exercises

Section A.

For each of the keystroke sequences listed below predict the value that will be shown in the displays of AL and AOS calculators respectively.

1. 5 $\boxed{+}$ 3 $\boxed{\times}$ 2 $\boxed{=}$

2. 10 $\boxed{-}$ 8 $\boxed{\div}$ 4 $\boxed{=}$

3. 18 $\boxed{\div}$ 3 $\boxed{\times}$ 2 $\boxed{=}$

4. 5 $\boxed{\times}$ 3 $\boxed{-}$ 4 $\boxed{\times}$ 2 $\boxed{=}$

5. 7 $\boxed{\times}$ 3 $\boxed{+}$ 5 $\boxed{\div}$ 2 $\boxed{=}$

6. 6 $\boxed{+}$ 12 $\boxed{\div}$ 3 $\boxed{\times}$ 2 $\boxed{=}$

7. 9 $\boxed{\times}$ 2 $\boxed{\div}$ 3 $\boxed{\times}$ 2 $\boxed{=}$

8. 9 $\boxed{\times}$ 2 $\boxed{\div}$ 3 $\boxed{\div}$ 2 $\boxed{=}$

9. 4 $\boxed{+}$ 3 $\boxed{\times}$ 2 $\boxed{=}$ $\boxed{\times}$ 5 $\boxed{=}$

10. 5 $\boxed{-}$ 3 $\boxed{\times}$ 2 $\boxed{=}$ $\boxed{\div}$ 2 $\boxed{=}$

11. 3 $\boxed{\div}$ 4 $\boxed{\times}$ 2 $\boxed{+}$ 1 $\boxed{\div}$ 2 $\boxed{=}$

12. 3 $\boxed{\times}$ 2 $\boxed{\div}$ 4 $\boxed{+}$ 5 $\boxed{\times}$ 4 $\boxed{\div}$ 2 $\boxed{-}$ 6 $\boxed{\times}$ 2 $\boxed{=}$

Section B.

For each of the following mathematical expressions, write keystroke sequences in both AL and AOS that will correctly compute the mathematical expression.

13. $5 + 3 \times 2$

14. $(5 + 3) \times 2$

15. $2 \times 3 - 5 \times 4$

16. $3 - 7 \times 4$

17. $\frac{2}{3} + \frac{4}{7}$

18. $\frac{3 \times 7 - 9}{2 \times 6}$

19. $2 \times 3 - 4 \times 5 + 7 \times 3$

20. $21/7 \times 4$

21. $\frac{1}{6} + \frac{2}{3} - \frac{3}{7}$

22. $3\ 1/8 + 4\ 3/5$
[Hint: $3\ 1/8 = 3 + 1 \div 8$]

23. $8 - 2\ 3/7$

24. If it were necessary for the answer to exercise 21 to be expressed in fractional form as opposed to decimal form, one might use the elementary principles of least common denominator to rewrite the expression as follows:

$$\frac{1}{6} + \frac{2}{3} - \frac{3}{7} = \frac{1(7) + 2(2)(7) - 3(6)}{6(7)}$$

Write the AOS and AL keystroke sequences necessary to compute the numerator of the second expression. What is the answer written in fractional form? Of course, it is probably faster to perform the computation above by mental arithmetic. However, can you change the numbers so that calculator computation might be preferable, using the calculator algorithm above?

Section C.

The keystroke sequences listed in the problems of section A above might be viewed as possible keystroke sequences to compute certain mathematical expressions. For example, the AL sequence in 1 computes the expression $(5 + 3)2$; The AOS sequence in 1 computes the expression $5 + (3)(2)$. For each of the sequences in numbers 2–12, write the mathematical expression that the particular keystroke sequence computes in AL and AOS.

2
Functions

IN contrast to binary operations, which are treated differently by different operating systems, unary operations or functions are performed in the same way on almost all calculators. Students should be introduced to functions very early in their mathematical training. Even elementary school students may enjoy playing a game that involves guessing the rule. Later in algebra courses functions are often represented by letters. For example:

$$f(x) = x^2 \qquad g(x) = \sqrt{x} \qquad h(x) = -x \qquad F(x) = 1/x$$

The four functions above, square, square root, additive inverse (or change sign), and multiplicative inverse (or reciprocal), will be the first four calculator functions discussed. In the notational scheme above, the function letter (f, g, h, or F) precedes the independent variable. Hence, using the above notation we might write $f(3) = 9$. We think of the function, f, as acting on the value 3 to produce the value 9. Many calculators have a squaring key, usually marked x^2. However, for all calculators, the function key is pushed *after* the independent variable is in the display.* This is consistent with right-hand function notation that is sometimes encountered when dealing with linear transformations (often in linear algebra texts) and might be written $(x)f = x^2$. Thus on a calculator with a squaring key, one would first push the number 3 and then the [x^2] key which would result in the number 9 being displayed. Notice that the [=] key is never used with

*A few programmable calculators (e.g., Sharp 5100) have adopted a mixed approach to functions which may be keyed either with right-hand or left-hand notation, depending on the function. For example, the square root function is keyed as left-hand notation but the square function is keyed as right-hand notation. The rationale appears to be that the function is keyed as it might be spoken. Usually one must use parentheses to indicate the variable to which the function is to be applied.

function keys, and the keystroke sequence above is the necessary one for AL, AOS, RPN, and arithmetic logic calculators. The following keystroke sequences and associated displays illustrate these four unary operations on all calculators:

Keystroke sequence	4 [x^2]	4 [$\sqrt{\ }$]	4 [+/−]	4 [1/x]
Calculator display	[16]	[2]	[-4]	[0.25]

When dealing with an expression containing both functions and binary operations, it is important to remember that the function will be performed on the result in the calculator display. The following examples will illustrate the calculation with unary operations or functions.

Mathematical expression	*AOS or AL keystrokes*	*Calculator display*
$3^2 + 5^2 - 7^2$	3	[3]
	[x^2]	[9]
	[+]	[9]
	5	[5]
	[x^2]	[25]
	[−]	[34]
	7	[7]
	[x^2]	[49]
	[=]	[-15]

The algorithm above is very useful in computing a sum of squares for a statistical calculation. One should be careful not to confuse a sum of squares with the square of a sum:

Mathematical expression	*AOS or AL keystrokes*	*Calculator display*
$(3 + 5)^2$	3	[3]
	[+]	[3]
	5	[5]
	[=]	[8]
	[x^2]	[64]

One should notice that the [=] key is used in both AL and AOS to complete the (binary) operation of addition before the unary operation of squaring is performed. Without the [=] key in the keystroke sequence above, the expression $3 + 5^2 = 28$ would be computed.

The square root key is employed similarly.

Mathematical expression	*AOS or AL keystrokes*	*Calculator display*
$\sqrt{4^2 + 3^2}$	4	[4]
	[x^2]	[16]
	[+]	[16]
	3	[3]
	[x^2]	[9]
	[=]	[25]
	[$\sqrt{\ }$]	[5]

Of course when there are binary operations of different hierarchies in an expression, the differences between AL and AOS keystroke sequences discussed in chapter 1 apply.

Mathematical expression	*AL keystrokes*	*AOS keystrokes*
$2(5)^2 + 3(4)^2$	2	2
	[×]	[×]
	5	5
	[x^2]	[x^2]
	[÷]	[+]
	3	3
	[+]	[×]
	4	4
	[x^2]	[x^2]
	[×]	[=]
	3	
	[=]	

The additive inverse (or change sign) key is indispensable to calculator operation. This key is usually represented either by [+/−] or by [CHS]. We shall employ the [+/−] designation. As indicated earlier in this chapter, the keystroke sequence of a number followed by [+/−] results in the negative of that number being displayed. All numbers are entered on the calculator as nonnegative decimals, so if we want to enter a negative number, we must first enter the positive number and then take its additive inverse. Unlike standard mathematical notation that uses the same symbol for the binary operation of subtraction and the unary function of additive inverse, the calculator has different keys to represent these two different concepts. Many novice calculator operators may try to use the subtraction key to enter negative numbers. If the negative number is the first used, then this may not lead to an error. For example, $0 - 5 = -5$.

However, consider the expression $5(-3)$. The necessary keystroke sequence for AL or AOS is shown at the right.

5
[×]
3
[+/−]
[=]

The use of the subtraction key (before the 3) would give the calculator conflicting instructions to perform both the binary operation of multiplication and the binary operation of subtraction. Most AL or AOS calculators will ignore the first binary operation and perform the second binary operation (subtraction, in this example).

Of course, subtraction can be defined in terms of addition and the additive inverse, $a - b = a + (\cdot b)$, which is useful in deriving calculator algorithms. For example, let us look at the possible ways we could compute an expression in AL that we previously encountered in chapter 1, example 7.

Mathematical expression	*AL* *Version 1*	*AL* *Version 2*	*AL* *Version 3*
$13 - 2 \times 4$	2 [×] 4 [=] [+/−] [+] 13 [=]	2 [+/−] [×] 4 [+] 13 [=]	2 [×] 4 [−] 13 [=] [+/−]

As mentioned in chapter 1, one of the problems here is that we would like to perform the multiplication first and then the subtraction, but subtraction is not commutative. In version 1 above we first compute the product, then take the additive inverse of the result, and finally perform the addition. Version 2 is only a slight variation on the first algorithm using the property that $-(ab) = (-a)(b)$. In version 3 we deliberately compute the difference in reverse order and then apply the mathematical property

$$(b - a) = a - b.$$

Combining all the functions thus far discussed in the chapter, we might have the following:

Mathematical expression	*AL*	*AOS*
$\dfrac{7 + \sqrt{7^2 - 4(3)(1)}}{2(3)}$	4 [+/−] [×] 3 [+] 7 [x^2] [=] [√] [+] 7 [÷] 2 [÷] 3 [=]	7 [x^2] [−] 4 [×] 3 [=] [√] [+] 7 [=] [÷] 2 [÷] 3 [=]

The expression above could arise in solving the quadratic equation $3x^2 - 7x + 1 = 0$ for the largest root by means of the quadratic formula:

$$x = \frac{-b \pm \sqrt{b^2 - 4ac}}{2a}$$

The multiplicative inverse (reciprocal) function operates the same for multiplication and division as the additive inverse (change sign) function does for addition and subtraction.

Mathematical expression	*AL*	*AOS*	*AL or AOS alternate*
$\dfrac{3(5)}{4 + 6}$	4 [+] 6 [÷] 3 [÷] 5 [=] [$1/x$]	4 [+] 6 [=] [÷] 3 [÷] 5 [=] [$1/x$]	4 [+] 6 [=] [$1/x$] [×] 3 [×] 5 [=]

Since we have previously discussed how to divide by a product, we use that technique in the first two solutions above by computing the reciprocal of the desired result. The last step in each of the first two solutions takes the reciprocal after all pending operations are completed (i.e., after the [=] key is pushed). For the alternative solution we choose to treat division by a quantity as multiplication by the reciprocal of that same quantity. In either case one should note that the reciprocal key operates on the number in the display, and so judicious use should be made of the [=] key. The sum of reciprocals is computed in a similar fashion to the sum of squares discussed previously.

Mathematical expression	*AL or AOS*
$\frac{1}{2} + \frac{1}{3} + \frac{1}{4} + \frac{1}{5} =$	2 [1/x] [+] 3 [1/x] [+] 4 [1/x] [+] 5 [1/x] [=]

All other functions on the calculator operate in analogous fashion. To use the logarithm function(s), display the argument and then depress the desired logarithmic key. To use trigonometric functions one must know whether the calculator is operating in degree mode, radian mode, or grad mode (1 revolution = 360° = 2 π radians = 400 grads). Since most calculators have at most three trigonometric functions (sine, cosine, and tangent), the reciprocal key may be used to advantage to obtain the other three trigonometric values using standard trigonometric identities. (See expression on next page.) Since there is no indication of degrees (in the mathematical expression), the variable ($3\pi/7$) is assumed to be expressed in radians. The radian mode on the calculator *must be selected* before the [cos] key is depressed. It may be less awkward to select the proper mode, radian in this example, *before* beginning the keystroke sequence. Notice the use of the [=] key to complete the pending binary operations. Since the secant function is the reciprocal of the cosine function, we compute the cosine and then

Mathematical expression	*AL or AOS*
sec (3 π / 7)	3 [×] [π] [÷] 7 [=] [cos] [1/x]

take the reciprocal of the result. Some novice users have a tendency to operate the [cos] and the [1/x] keys in reverse order.

A function g is said to be the inverse of a function f provided that $g(f(x)) = x$, or in right-hand notation, $((x)f)g = x$ for all x. The square function is the inverse of the square root function over the domain of the nonnegative real numbers (where the square root function is defined). The additive inverse and the multiplicative inverse are each their own inverses. To demonstrate this, experiment with keystroke sequences similar to those below varying the initial parameter.

4	4	4	4
[x^2]	[√‾]	[+/−]	[1/x]
[√‾]	[x^2]	[+/−]	[1/x]

In each example, the original number should appear in the display after the keystroke sequence of a function followed by its inverse (the number 4 in the example above).

On many calculators, the inverse of some functions are found by use of an [INV] key (for inverse function). Frequently the square root is calculated in this manner, by first pressing [INV] and then the square function key ([x^2]). The inverses of the trigonometric or logarithmic functions can be similarly calculated on most calculators. However, it should be noted that in taking the inverse of trigonometric functions, only the principal inverse is displayed. The following examples (in degree mode) illustrate the outcomes:

Keystroke sequence	150 [sin] [INV] [sin]	210 [sin] [INV] [sin]	330 [sin] [INV] [sin]	150 [cos] [INV] [cos]	210 [cos] [INV] [cos]	330 [cos] [INV] [cos]
Display	30	-30	-30	150	150	30

The range of the $\sin^{-1}$ function is $[-90°, 90°]$; the range of the $\cos^{-1}$ function is $[0°, 180°]$. Only within the ranges of the inverse trigonometric functions can one expect the composition of the trigonometric function and its inverse to yield the identity.

TABLE OF COMMON CALCULATOR FUNCTIONS AND THEIR INVERSES			
Function	*Inverse*	*Function*	*Inverse*
[+/−]	[+/−]	[sin]	[Arc sin] or [$\sin^{-1}$]
[$1/x$]	[$1/x$]	[cos]	[Arc cos] or [$\cos^{-1}$]
[x^2]	[$\sqrt{\ }$]	[tan]	[Arc tan] or [$\tan^{-1}$]
[$\log x$]	[10^x]	[$\ln x$]	[$\exp x$] or [e^x]

The following example illustrates the computation of exponential expressions using [INV] with the logarithmic function key on calculators that do not have an exponential key [y^x].

Mathematical expression	*AL or AOS*
$3^{1.5}$	3 [log] [×] 1.5 [=] [INV] } equivalently, [10^x] [log] }

The solution above is based on the basic logarithmic property that $\log a^r = r(\log a)$. Taking the inverse of both sides gives

$$a^r = \log^{-1}(r(\log a)).$$

Of course, the natural logarithm, ln, and the exponential function, e^x or exp, may be used analogously to evaluate $3^{1.5}$ or similar exponential expressions. The only possible drawback is that the base of the exponential expression must be positive, since the domain of the logarithmic function is the positive reals. However, many calculators with exponential binary operation keys will also abort attempted expressions like $(-2)^3$, performed with the [y^x] key even though such expressions are well defined, mathematically.

Review

1. Describe four functions (unary operators) often found on calculators.
2. How does the execution of unary operations on a calculator differ from the execution of binary operations on AL or AOS calculators?
3. What is the inverse of each of the functions in 1?
4. Differentiate between subtraction and the additive inverse. How does standard algebraic notation disguise these differences? Why is it important that one differentiate between these two concepts when operating the calculator?

Exercises

Section A.

For each of the keystroke sequences listed below, predict the value that will be shown in the displays of both AL and AOS calculators.

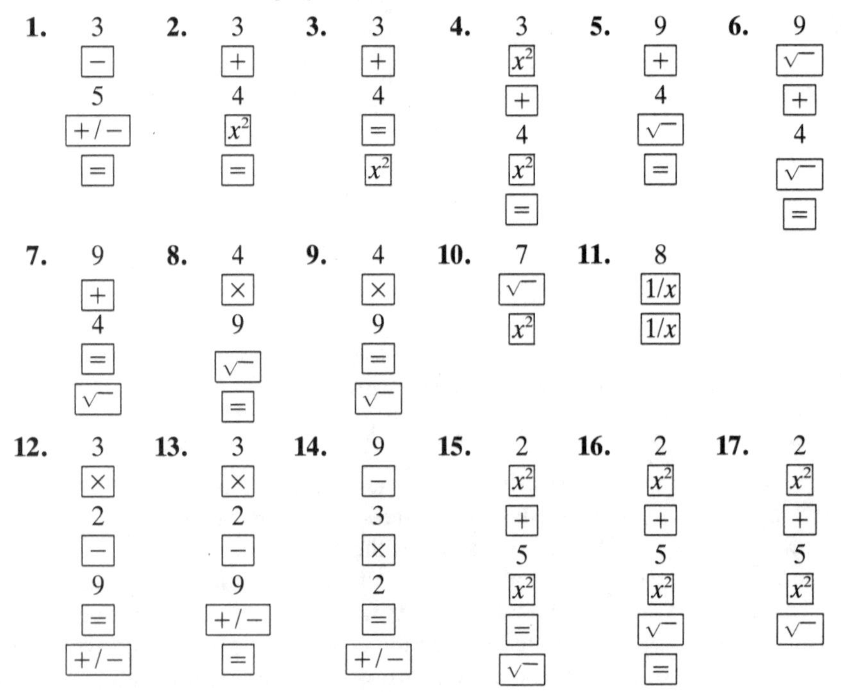

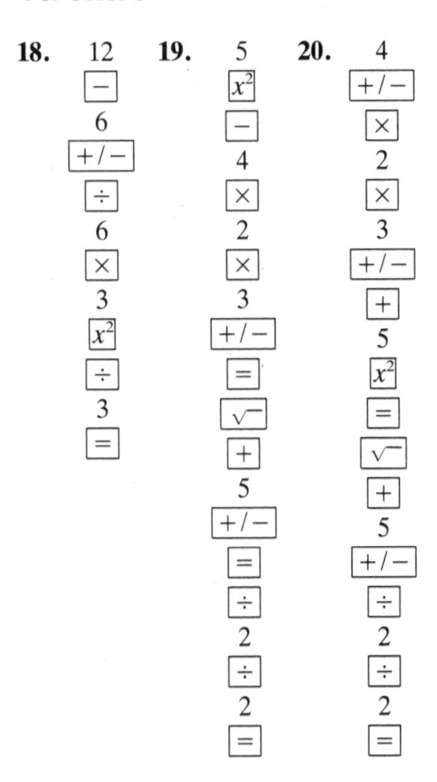

Section B.

List the keystroke sequences (in both AL and AOS) necessary to solve each of the following quadratic equations using the quadratic formula.

21. $2x^2 + 11x - 21 = 0$

22. $6x^2 + 17x + 5 = 0$

23. $4x^2 + 12x + 9 = 0$

24. $2x^2 + 5x - 4 = 0$

25. $x^2 - 3x + 1 = 0$

26. $2x^2 + 5x + 4 = 0$

27. $x^2 - x + 1 = 0$

28. $5.39\,x^2 + 3.1x - 0.27 = 0$

29. Use your own calculator to solve those equations above that have real solutions. Most calculators cannot be used to compute complex numbers. How can equations 26 and 27 above be solved using the calculator? Does the calculator give exact or approximate solutions for each of the above equations?

Section C.

If any number a is originally in the display of the calculator, decide whether the following keystroke sequences yield the same or different results. Tell which numbers yield different results and whether the operating system makes any difference. [Hint: Remember that the number a may be negative or zero.]

30. a [x^2] [+/−] and a [+/−] [x^2]

31. a [x^2] [$1/x$] and a [$1/x$] [x^2]

32. a [$1/x$] [+/−] and a [+/−] [$1/x$]

33. a [+/−] [$\sqrt{\ }$] and a [$\sqrt{\ }$] [+/−]

34. a [$\sqrt{\ }$] [$1/x$] and a [$1/x$] [$\sqrt{\ }$]

35. a [$\sqrt{\ }$] [x^2] and a [x^2] [$\sqrt{\ }$]

Section D.

Let a be any number on the face of the calculator. Let b be any number that can be entered from the keyboard (i.e., $b \geq 0$). Decide whether each of the following keystroke sequences yields the same or different results. How do the values of a and b change your conclusions? [Of course, the use of the [=] key implies the use of an AL or AOS calculator.]

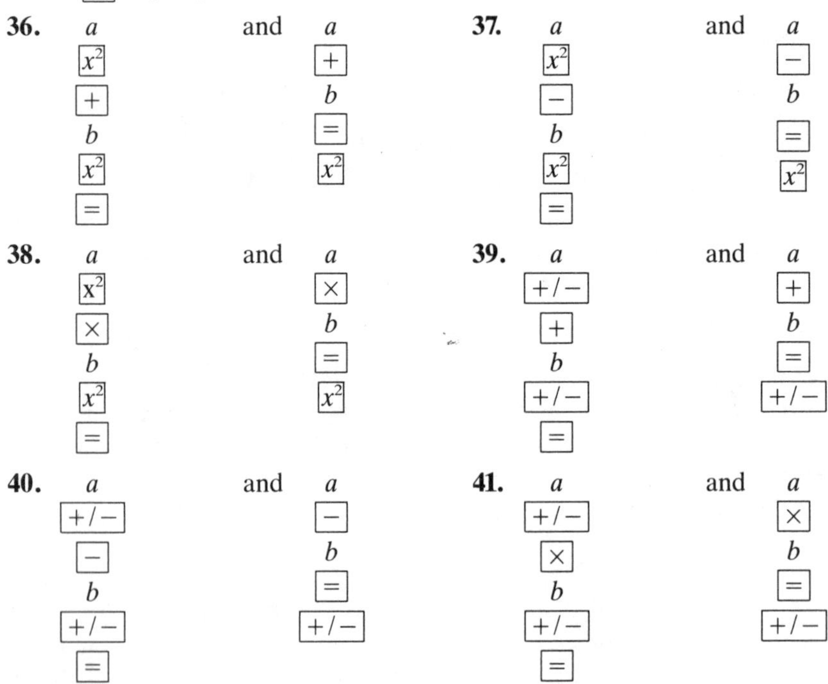

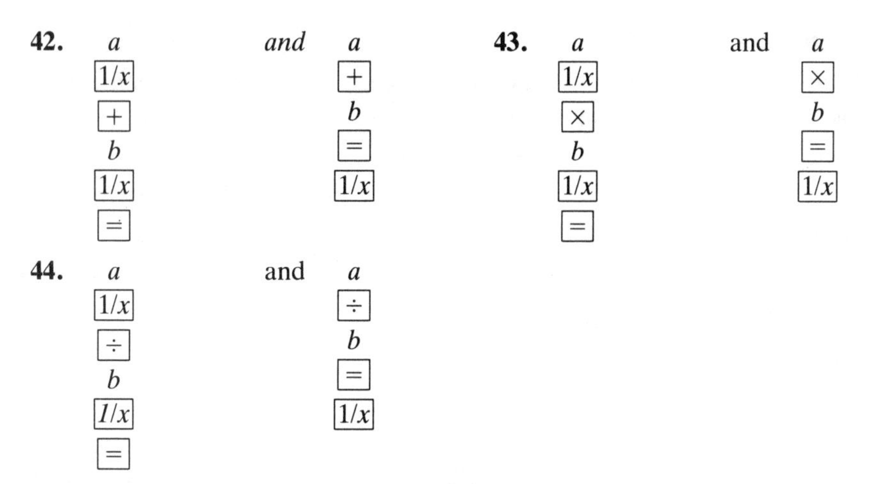

45. In general if # is a function and * is an operation, then the function # is distributive over the operation * if the two keystroke sequences at the right yield the same results for all values of a and b. Which functions that we have discussed in this chapter are distributive over any of the four basic operations?

a	and	a
#		*
*		b
b		=
#		#
=		

Section E.

An iteration may be described as the repeated use of a sequence of functions or operations. Predict the outcomes of each of the following iterations. In each case try the iteration with different values of a. In particular, in 48 and 49 try values both larger and smaller than 1. In 50 try letting a = 6 and a = 12. Record the results after each iteration, and if the sequence is converging, predict the limit of convergence.

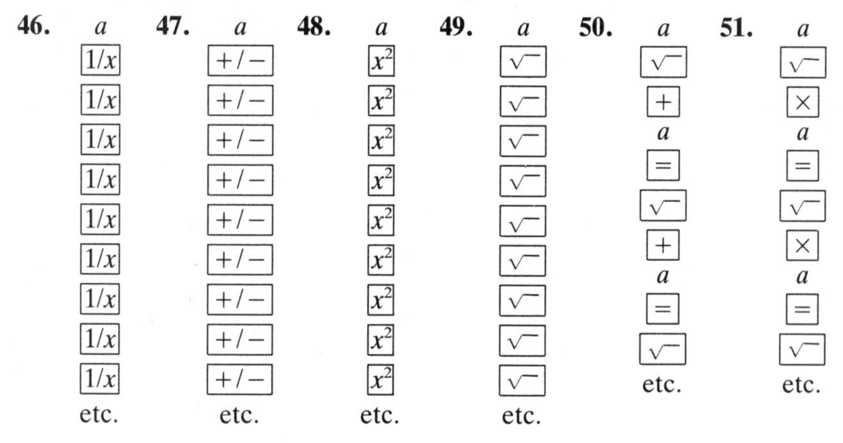

Section F.

Write the AL and AOS keystroke sequences necessary to compute each of the following mathematical expressions.

52. $3 / 17 - 4 / 5$

53. $3^2 + 4^2 + 7^2$

54. $15 - 3 \times 2$

55. $\dfrac{2(9)}{5 + 3}$

56. $\cos(30°)$

57. $\sec(30°)$

58. $\sin(3\pi/4)$

59. $\csc(3\pi/4)$

60. $\sin^{-1}(0.3)$

61. $\csc^{-1}(3.5)$

3 Addressable Memory

AN addressable memory is a valuable tool on any calculator. It may be used to store a number that will be needed later in a calculation. Two important uses are (1) to store intermediate results in lengthy calculations and (2) to store constants that may be used several times within the same calculation. The number in the addressable memory is stored electronically within the machine and is usually not visible to the user. However, on many calculators, a small light or the letter *m* will appear in the display if a nonzero number is stored in the memory. No calculator will have all of the memory keys that will be discussed.

The table on the next page lists the most important uses of the memory and the keys that perform these functions. It is important to understand that each row of the table is independent and operates on the original (x, y) display/memory pair and not on the results of the previous row.

The *clear memory* key (often denoted by [CM] or [MC]) stores the number zero in the memory. It is probably more prudent to use it just before you wish to clear the memory rather than relying on the habit of clearing the memory before each calculation.

The *recall memory* key (often denoted by [RCL], [RM], [$x \leftarrow$ m], or [M out]) duplicates or copies the value in the memory into the display. It is important to remember that the value in the memory remains in the memory and the previous value in the display is lost. For example, if 5 was in the display and 3 was in the memory, after the recall ([RCL]) key was depressed, the number 3 would be stored both in the memory and in the display. Although most of the addressable memory keys are optional or nonstrategic, the *recall memory button* is essential to display the accumulated value in the memory or to reproduce memory results in a calculation.

The *add to memory* key (often denoted by [M+] or [SUM]) adds the

TABLE OF MEMORY FUNCTIONS			
Task	Common keys	Display	Memory
Original Contents		x	y
Clear Memory	[CM] or [MC]	x	0
Recall	[RCL] or [x←m] or [MR] or [M out]	y	y
Add to Memory	[M+] or [SUM]	x	$y + x$
Subtract from Memory	[M−] or [DIFF] or [INV SUM]	x	$y - x$
Store	[STO] or [x→m] or [M in]	x	x
Exchange	[EXC] or [x↔m]	y	x
Multiply times Memory	[PROD]	x	yx
Divide into Memory	[QUO] or [INV PROD]	x	y/x

contents of the display to the value in the memory. Usually the current value in the display is added to the value in the memory so that the keystroke sequence shown at the right produces the number 8 in the display. However, on calculators with a so-called live memory, operations pending in the display are completed before any memory functions are performed. With such a calculator, the result displayed after the keystroke sequence above will be 18. It is important to determine which type of memory operation your calculator supports.

3
[CM]
[M+]
[x]
5
[M+]
[=]
[RCL]

For the purposes of this discussion we will assume that the memory is *not* a live one. If your calculator has a live memory, many of the algorithms that we will develop will have to be modified. Thus if the value 3 is in the memory and 5 is in the display, then after the [M+] (or [SUM]) key is pressed, the value 8 will be in the memory and the value 5 will remain in the display no matter what the previous operations may have been.

Example: Suppose that we wish to compute both the sum of squares and the sum of a given data set. For example, we wish to compute the sum $3 + 5 + 8 = ?$ and the sum of squares $3^2 + 5^2 + 8^2 = ?$

Solution: Entering the data set twice is a duplicated effort. A more efficient algorithm is to compute the sum in the memory while simultaneously computing the sum of squares in the display, as shown in the keystroke sequence at the right. After recording the result shown in the display, we would use the [RCL] key to display the value stored in the memory.

AL or AOS

3
[CM]
[M+]
[x^2]
[+]
5
[M+]
[x^2]
[+]
8
[M+]
[x^2]
[=]

The *subtract from memory* key ([M−] or [DIFF]) operates for subtraction as the *add to memory* key operates for addition. Since subtraction is noncommutative, it is important to remember that the value in the display is subtracted from the value in the memory and the result is stored in the memory. For example, if the number 5 was in the display and the number 3 was in the memory, then after activating the [M−] (or [DIFF]) key the number −2 would be stored in the memory and the number 5 would remain in the display.

The *store* key ([STO] or [x→m] or [M in]) copies the number in the display into the memory. *The store key is equivalent to the sequence* [CM], [M+]. If the number 5 was originally in the display and the number 3 was originally in the memory, then after activating the [STO] key the number 5 would be saved both in the memory and in the display. Since the store key "erases" the previous value stored in the memory, most calculators that include a store function do not also have a clear memory key.

The *exchange* key ([EXC] or [x↔m]) exchanges the contents of the display with the contents of the memory. For example, if the number 5 was originally in the display and the number 3 was originally in the memory, then after activating the [EXC] key the number 5 would be in the memory and the number 3 would be in the display. The exchange function is extremely useful for reminding oneself of the value in the memory without losing the value in the display. Clearly, activating the [EXC] key twice restores the calculator to its original state.

The *multiply times memory* key ([PROD]) multiplies the number in the memory by the number in the display. If the number 5 is in the display and the number 3 is in the memory, then after activating the [PROD] key the number 15 is in the memory, whereas 5 is retained in the display.

The *divide into memory* key ([QUO] or [INV PROD]) divides the number in the memory by the number in the display and stores the result in the memory. If the number 5 is in the display and the number 10 is in the memory, then after activating the [QUO] key, the number 2 is stored in the memory and 5 is retained in the display. Since division is noncommutative, it is important to remember that the contents of the memory are divided by the contents of the display and not vice versa.

Some calculators have more than one addressable memory. If there are only two, they will sometimes be listed as different keys on the face of the calculator. More often, though, the different memories are accessed by first pushing the memory function key and then the address, which is usually a numeral. For example, if there are ten memories, their usual addresses are the digits 0 through 9. If there are 100 memories, their addresses are two-digit numbers 00, 01, 02, . . ., 99. In the latter case it is important to remember to enter the two-digit address, for example 02, and not the single digit 2. For calculators with more than one addressable memory, one must remember to use the desired address each time *any* memory key is activated.

There is a second type of memory, called an operating memory or operating register, that is used by the calculator to store pending numbers and operations. For example, if an AL or AOS calculator is to perform the sum 3 + 5, the calculator must store the value 3 and the operation + in an operating register until after the value 5 is entered. In general, the values that the calculator retains in operating registers are not directly available to the user.

At times one may wish to clear all numbers from the calculator and start over with a clean slate. The easiest way to do this on many calculators is to turn the calculator off and then turn it on again. However, some calculators are constructed with a so-called nonvolatile memory, which retains its contents even when turned off. The obvious advantage of the nonvolatile memory is that one cannot inadvertently lose desired stored values. However, unless the calculator has an *all-clear* button, the only way to clear all memories is by directly addressing each one.

For most purposes it is not necessary to clear the memory or the display. As mentioned above, the [STO] key clears the memory of previous values. On AL and AOS calculators, pressing the [=] key clears all operating registers, leaving only the result in the display (and, of course, any values that may be stored in addressable memories). When a new calculation is initiated by entering the first numerical value in the new expression, that new number replaces the one previously stored in the display and the calculator is effectively cleared of the previous calculation. However, if one intends to use the result from a previous calculation as input to a new expression, one can start with the result from the previous problem in the display and *enter the operation or function* needed for the new expression. Thus, the number from the previous calculation will be treated as if it were entered as the first numerical entry of the new expression. Not reentering intermediate results clearly reduces the opportunities for copying errors and saves time and keystrokes. Furthermore, since many calculators carry more digits internally than are shown in the display, such additional precision would be lost if intermediate results were reentered.

An invaluable aid to calculator manipulation is the *clear entry* key, which clears the last numerical entry but not other operating registers. If one miskeys a number in a lengthy calculation, by using the clear entry key one can correct this mistake without reentering the entire keystroke sequence. Many manufacturers use a single key as both a *clear entry* key and as a *clear* key. The [CE/C] key is usually depressed once to clear only the most recent entry in the display and pressed a second time to clear all operating registers and the display.

Review

1. Name two purposes for using an addressable memory.
2. The word *output* is often used when speaking of computers (or other electronic information-storing machines) to indicate information that is sent *from* the machine environment *to* the human environment. How do we get output from the addressable memory of the calculator?
3. The word *input* is often used when speaking of computers (or other electronic information-storing machines) to indicate information that is sent *from* the human environment *to* the machine environment. There are several means of getting input to the addressable memory of the calculator. Discuss them, explaining how they differ.
4. Distinguish between an addressable memory and an operating memory.
5. Distinguish between the *clear, all clear,* and *clear entry* functions. Describe different situations for which each clearing function would be appropriate.

Exercises

1. Fill in the following table giving the result in both the display and the memory after each calculator function is performed on the original (3,12) pair.

Calculator key used	*Display*	*Memory*
	3	12
CM		
M+		
M−		
STO		
RCL		
PROD		
QUO		
EXC		
CE		
+/−		

2. Repeat exercise 1, starting with the original pair (15,3).
3. List the keystroke sequence used to enter the pair (3,12) into the display and memory.
4. List the keystroke sequence needed to enter the pair $(5,-3)$ into the display and memory.
5. In chapter 2, p. 23, #50, a keystroke sequence was used to calculate $\sqrt{a+\sqrt{a+\sqrt{a+\sqrt{a+\sqrt{a+}}}}}\ldots$. When a is a single-digit number, it is not difficult to reenter the value a each time it is needed. Describe an alternative keystroke sequence that one might want to use if a is a multidigit number. Use your calculator to predict the convergence of the sequence above when $a = 7.3125$.

4
Exponentiation and Additional AL and AOS Strategies

EXPONENTIATION is a binary operation that has been previously alluded to but not discussed fully. Many calculators have a [y^x] key. The keystroke sequence at the right might be used to compute 2^3.

2
[y^x]
3
[=]

If the treatment of exponentiation was to be consistent with that of other binary operations, the AOS calculator would execute exponentiation prior to all other operations, whereas the AL calculator would perform exponentiation on the result of all pending operations. This would give the following display results for the given keystroke sequences:

	A	*B*	*C*	*D*	*E*
	5 [−] 2 [y^x] 3 [=]	3 [×] 2 [y^x] 3 [=]	2 [+/−] [y^x] 3 [=]	2 [y^x] 3 [+/−] [=]	2 [y^x] 3 [=] [+/−]
AOS display	-3	24	-8	0.125	-8
AL display	27	216	-8	0.125	-8

However, these predicted results for *A* and *B* above occur on *some but not all* calculators, although very few calculators will give the result shown in *C* above. Although $(-2)^3 = -8$, most calculators will not perform exponentiation on negative numbers. (See also the discussion of logarithms, on pp. 1–2) Of course, $(-1)^{1/2} = (-1)^{0.5} = \sqrt{-1}$, is an imaginary number, which is not in the domain of most calculators. One possible way to compute a^b for $a < 0$ is $a^b = (-1)^b \, |a|^b$. $(-1)^b$ may be readily computed without the aid of a calculator for both integral and fractional values of b.

Nearly all calculators with the [y^x] key will give the results listed in D and E above. However, since the keystrokes are quite similar but yield very different results, one should be careful to distinguish between the two keystroke sequences. Keystroke sequence D computes the expression 2^{-3}; sequence E computes the expression -2^3.

The [y^x] key and the [x^2] and [$\sqrt{\ }$] keys provide two different ways of thinking about expressions such as 3^2 and $\sqrt{3}$: as the squaring (square root) *function* or as the *binary operation*, y^x combining two real numbers into a third real number.

3	or	3	3	or	3
[x^2]		[y^x]	[$\sqrt{\ }$]		[y^x]
		2			0.5
		[=]			[=]

The [x^2] key will usually give the expected result when the base (3 in the example above) is negative. Hence $(-3)^2$ may be calculated using the [x^2] key, even though $(-3)^2$ usually cannot be computed using the [y^x] key. The [x^2] and [$\sqrt{\ }$] keys obviously employ fewer keystroke sequences. However, since most calculators do not have built-in cubing or other higher-powered or decimal exponentiation functions, the [y^x] key is applicable in a greater variety of calculations.

The following examples illustrate possible algorithms for calculating various exponential expressions with AL or AOS:

Mathematical expression	*AL*	*AOS*
$3(4^3) - 2^3$	2	3
	[y^x]	[×]
	3	4
	[=]	[y^x]
	[STO]	3
	4	[−]
	[y^x]	2
	3	[y^x]
	[×]	3
	3	[=]
	[−]	
	[RCL]	
	[=]	

The AOS sequence above takes advantage of the hierarchical execution of binary operations by the machine. The AL solution requires *the user* to perform operations in hierarchical order and to store intermediate results in the calculation.

Mathematical expression	*AL*	*AOS*
$[3(4) - 2]^3$	3 [×] 4 [−] 2 [y^x] 3 [=]	3 [×] 4 [−] 2 [=] [y^x] 3 [=]

In the AOS solution the [=] key must be used to complete pending operations before [y^x] key is used. If the AL calculator is calculating all binary operations sequentially, then pressing the [y^x] key completes pending operations (or closes the parentheses in the mathematical expression). If your calculator does not give the correct display results of 1000, its operating system may treat exponentiation differently from other binary operations. In such cases you will have to experiment with various probable combinations to determine what will work on your calculator.

3
[×]
[=]
[=]
[=]

For calculators without exponentiation keys one may employ logarithms as previously discussed in chapter 2. Also, some calculators allow calculation of integral-valued exponents by repeated multiplication. The keystroke sequence at the right shows how to compute 3^4 on these calculators. One should notice that there is one less [=] sign in the keystroke sequence than in the exponent. The use of repeated multiplication is further illustrated in the calculation of the two mathematical expressions on the next page, which must be carefully distinguished. In the right-hand column, the multiplication after the 6 completes pending operations, so the suceeding two [=] keystrokes cube the entire quantity.

Repeated operation by a constant may be performed on many calculators by storing the number and the operation. Conceptually, the user is defining a new function. For example, the user might enter the new function that multiplies each independent variable from a list by 1.15 (to perform 15 percent markup). The user stores the number 1.15 and the operation of

Mathematical expression	*AL with repeated multiplication*	*Mathematical expression*	*AL with repeated multiplication*
$6 - 4(2^3)$	2 [×] [=] [=] [×] 4 [+/−] [+] 6 [=]	$(6 - 4(2))^3$	4 [×] 2 [+/−] [+] 6 [×] [=] [=]

multiplication. As long as no other binary operations are entered, the [=] key will operate as the newly defined function key. When a number on the list is entered followed by the [=] key, it is automatically multiplied by 1.15.

The way in which the constant number and operation are stored (for these user-defined functions) differs among calculators. On some machines (usually AL) the constant and the operation are entered first, as in the keystroke sequence at the right. The value in the display after the first [=] key is pushed is 4.5; after the second [=] key, the number 10.5 appears on the display. On other machines, or with a different operation, such as division, the number retained as the constant will be the second number entered (the 3 in the example above). Actually, the exponentiation by repeated multiplication discussed earlier is simply storing the base and the operation of multiplication. Each time the [=] key is activated, the value in the display is multiplied by the stored base.

1.5
[×]
3
[=]
7
[=]

On other machines (frequently AOS), the constant number and operation for this type of user-defined function are stored by pressing a special key, usually designated by *k* or *const.* In the keystroke sequence shown at the right, the number 3.1415927 (approximately π) and the operation of division are stored by pressing the key [*k*]. The value in the display after the first [=] key is pushed is $5 \div \pi = 1.5915494$; after the second [=] key, the value $7 \div \pi = 2.2281692$ appears on the display. Before actually using repeated operations to calculate unknown results, one should experiment with simple data to be sure of how a given calculator stores and controls a particular binary operation. If none of the techniques

3.1415927
[÷]
[*k*]
5
[=]
7
[=]

suggested above works with a given calculator you may need to consult the owner's manual.

The remainder of this chapter will outline some typical keystroke sequence algorithms for algebraic logic (AL) or algebraic operating systems (AOS) calculators. Many of these algorithms will use built-in calculator functions or memory functions. Although we could not, of course, illustrate every type of problem that you might encounter, familiarity with these procedures will help you become resourceful at creating your own algorithms for other types of expressions.

Example 1: *Sum or difference of products [ab + cd − ef . . .]*

Mathematical expression	*AL Version 1*	*AL Version 2*	*AL Version 3*	*AL Version 4*	*AL Version 5*
13 − 3(2)	13 [÷] 2 [−] 3 [×] 2 [=]	3 [×] 2 [−] 13 [=] [+/−]	3 [×] 2 [+/−] [+] 13 [=]	3 [×] 2 [=] [STO] 13 [−] [RCL] [=]	13 [STO] 3 [×] 2 [=] [M−] [RCL]

The AOS keystroke sequence for this expression, which was presented previously, is straightforward and uses the mathematical hierarchy of execution of operations in the AOS calculator. All the AL solutions above, except for AL versions 4 and 5, were previously discussed. In version 4 one should note that the number being subtracted (subtrahend), not the minuend, is stored in the memory. In version 5 the memory is used as an accumulator, with successive products being added to or subtracted from it. Versions 2 and 3 are obviously the shortest keystroke sequences for this expression. However, for the more general form of the expression in which the minuend is also a product (e.g., 5(3) − 3(2)), versions 2 and 3 are not appropriate. This latter example could easily be accomplished using the algorithms represented in versions 1, 4, or 5.

An often-asked question is why not use the AOS machine, which has fewer keystrokes and may require fewer thought processes to use correctly. A pragmatic answer is that the algebraic logic machine exists and a calculator-literate person may need to know how to use the AL machine. The fact that the AOS machine performs operations in hierarchical order and may require fewer thought processes (to use the machine) is detrimental at some stages of mathematical development when we want to teach the hierarchy of operations. Moreover, the mathematically trained person may find it more natural to continue to control multiplications manually prior to executing additions and subtractions. Although the AOS machine requires fewer keystrokes to complete the solution to this problem, for other types of expressions the reverse is true.

Example 2: *Quotient of two sums* $\left[\frac{ab + cd - ef + \ldots}{rs + tu - vw + \ldots}\right]$

Mathematical expression	*AOS*	*AL*
$\frac{5(3) + 2(7)}{4(7) - 6(3)}$	4 [×] 7 [−] 6 [×] 3 [=] [STO] 5 [×] 3 [+] 2 [×] 7 [=] [÷] [RCL] [=]	6 [×] 3 [=] [STO] or [CM], [M+] 4 [×] 7 [−] [RCL] [=] [STO] or [CM], [M+] 5 [×] 3 [÷] 7 [+] 2 [×] 7 [÷] [RCL] [=]

The algorithm for both AOS and AL is to (1) compute and store the expression in the denominator, (2) compute the expression in the numerator, and (3) divide the value in the numerator (display) by the value stored in the memory. The AOS user should be careful to use the [=] key after calculating the numerator before dividing by the stored value of the denominator. Neglecting to do so would result, instead, in the calculation of the expression

$$5(3) + \frac{2(7)}{4(7) - 6(3)}.$$

The AL user might attempt to store all intermediate products before summing them, as was done in computing the value in the denominator. However, note that we could not use that same strategy in the numerator, since the memory was already being used to store the denominator (assuming only one memory). Hence, we reverted to the original algorithm developed in chapter 1 (AL version 1 in the previous example) for computing sums of products. In AL the [=] key is not needed to complete pending operations after the numerator is keyed in, since the division operation will automatically complete operations before dividing by the result in the memory.

Of course if the denominator (or numerator) has only one term, then we may revert to the early algorithm of dividing the result in the numerator by each factor of the denominator. (See pp. 4, 9, and 17.)

Example 3: *Product of several exponential expressions* $[(ab + cd)^m (ef + gh)^n \ldots]$

Mathematical expression	*AL*	*AOS*
$(5 - 6/7)^2 (6\sqrt{7} - 7/5)^4$	7 [÷] 5 [=] [STO] 7 [√‾] [×] 6 [−] [RCL] [y^x] 4 [=] [STO] 6	6 [×] 7 [√‾] [−] 7 [÷] 5 [=] [y^x] 4 [=] [STO] 5 [−] 6

Continued on next page

(continued)

	AL	*AOS*
	÷ 7 +/− + 5 = x^2 × RCL =	÷ 7 = x^2 × RCL =

The strategy is to compute each successive factor, store it, and then multiply the next factor by the stored result. The solutions above use the alternatives of the y^x and x^2 keys.

Example 4: *The sum of powers $[a^m + b^n + \ldots]$ versus the power of a sum $(a + b + \ldots)^m$*

Mathematical expression	*AL*	*AOS*	*Mathematical expression*	*AL*	*AOS*
$3(5) - 4^3$	4 y^x 3 = STO 3 × 5 − RCL =	3 × 5 − 4 y^x 3 =	$[3(5) - 4]^3$	3 × 5 − 4 y^x 3 =	3 × 5 − 4 = y^x 3 =

The importance of the distinction between operating systems is illustrated in these two examples. The AOS keystroke sequence given for the expression on the left is the same as the AL sequence for the expression on the right. This example shows that it is not always true that AOS keystroke sequences are shorter than their AL counterparts.

Example 5: *Use of the reciprocal function in calculating quotients*

Mathematical expression	*AL*	*AOS*
$\dfrac{(2t - z)^3 + t^3}{(2w - z + t)^3}$	2	2
	[×]	[×]
	5	5
where $w = 3$	[−]	[−]
$t = 5$	4	4
$z = 4$	[y^x]	[=]
	3	[y^x]
	[=]	3
	[STO]	[+]
	5	5
	[y^x]	[y^x]
	3	3
	[+]	[=]
	[RCL]	[STO]
	[=]	2
	[STO]	[×]
	2	3
	[×]	[−]
	3	4
	[−]	[+]
	4	5
	[+]	[=]
	5	[y^x]
	[y^x]	3
	3	[=]
	[÷]	[EXC]
	[RCL]	[÷]
	[=]	[RCL]
	[$1/x$]	[=]

Since the AL solution above requires the use of the memory to store intermediate results to calculate the numerator, we cannot use it in first calculating the denominator as outlined in the preceding examples. Hence, we have chosen first to calculate and store the numerator. We then compute the denominator and divide by the stored numerator, giving the reciprocal of the desired result. The final step is to use the multiplicative inverse function to calculate the desired result (i.e., the reciprocal of the reciprocal). The AOS solution above shows an alternative of using the [EXC] key to swap contents of the display and memory when one has inadvertently or otherwise stored values in inconvenient cells. Of course, since the AOS solution does not require the use of the memory to store intermediate results in either the denominator or numerator, a more efficient algorithm would first calculate and store the denominator, as was suggested in previous examples.

Example 6: *Use of parentheses*

So far we have resisted the temptation to use *parentheses* keys. Sometimes a saving in the number of needed keystrokes can be made by using the [=] key (one keystroke) instead of the parenthesis pair (two keystrokes). However, more important, some expressions cannot be calculated in a left-to-right keystroke sequence, as illustrated in the following three examples:

$$\text{(A)}\ 3 + 4 \times 5^{(11 - 2 \times 6)}$$

$$\text{(B)}\ 4 + 5(6 - 7 \times 2^3)$$

$$\text{(C)}\ 3(7 - (4 + (6\,(5 - 3 \times 2) - 8)))$$

If one keys in each of the expressions above as written left to right, including the parentheses keys, nearly all AOS machines will display an error, indicating an overflow of the number of allowed pending operations. *Machines have finite capabilities.* In the second expression above, the first addition is a pending operation until the multiplication (by 5) is performed. The first multiplication (by 5) is a pending operation until the parentheses are closed. The subtraction in the parentheses is a pending operation until the product (7×2^3) is computed. The multiplication (by 7) is a pending operation until the exponentiation (by 3) is completed. There are thus four pending operations and one set of parentheses in the expression, so if the calculator supports five or fewer pending operations and parentheses, it will overflow. It should be noted that closing parentheses can also be accomplished using the [=] key. As shown below, the three expressions above are easily calculated in either AOS or AL without resorting to the use of parentheses keys.

(A)

Mathematical expression	*AL*	*AOS*
$3 + 4 \times 5^{(11 - 2 \times 6)}$	2 +/− × 6 + 11 = STO 5 y^x RCL × 4 + 3 =	11 − 2 × 6 = STO 3 + 4 × 5 y^x RCL =

(B)

Mathematical expression	*AL*	*AOS*
$4 + 5(6 - 7 \times 2^3)$	2 y^x 3 × 7 +/− + 6 × 5 + 4 =	6 − 7 × 2 y^x 3 = × 5 + 4 =

(C)

Mathematical expression	*AL*	*AOS*
3(7 − (4 + (6 (5 − 3 × 2) − 8)))	3 [+/−] [×] 2 [+] 5 [×] 6 [−] 8 [+] 4 [=] [+/−] [+] 7 [×] 3 [=]	5 [−] 3 [×] 2 [=] [×] 6 [−] 8 [=] (optional) [+] 4 [=] [+/−] [+] 7 [=] [×] 3 [=]

One should not conclude that the parentheses keys should never be used, but rather that one should make judicious use of such keys and consider alternatives to mechanical left-right rote entry of mathematical expressions when using the AOS calculator.

In the keystroke sequence for the expression on the following page, it is advantageous to use parentheses. In this expression, $w = 3.7925$ was stored for using wherever needed in the calculation. Storing such a constant reduces both the chance of error and the number of keystrokes needed, particularly if the constant has many digits. If there is only one addressable memory, then it cannot also store an intermediate value in the calculation as was done previously in storing the denominator. In the solution above, the numerator was first calculated, and all pending operations were completed by using the [=] key. Then, the pair of parentheses was used to group the three terms of the denominator. It is important to remember that in a mathematical expression, the horizontal bar is a grouping symbol along with parentheses, brackets, or braces.

Mathematical expression	*AOS*
Evaluate $\dfrac{2w + 5w^2 + 4}{3w^2 - w + 1}$ for $w = 3.7925$	3.7925 [STO] [×] 2 [+] 5 [×] [RCL] [x^2] [+] 4 [=] [÷] [(] 3 [×] [RCL] [x^2] [−] [RCL] [+] 1 [)] [=]

The concluding sections of this chapter will outline the use of the calculator for two important applications: (1) finding the present value of an annuity and (2) finding the standard deviation of a set of values.

1. *Present value of annuity.* Consider an annuity that pays R dollars a period for n periods at a constant interest rate of i in each period. Denoting the present value of the annuity as A, we get

$$A = R \cdot \frac{1 - (1 + i)^{-n}}{i}, \text{ or } R = \frac{A \cdot i}{1 - (1 + i)^{-n}}$$

Example: Suppose we wish to borrow \$15 000 for 10 years to be paid off through monthly payments at 13 percent a year. We wish to know what the monthly payments, R, will be for this loan. The present value of the loan,

A, is 15 000; the number of pay periods, n = (10 years) (12 months/year) = 120; the interest rate, i = (0.13 per year)/(12 months/year) = 0.13/12. We need to calculate as follows:

	AL	*AOS*
$R = \dfrac{(15\,000)\,(0.13/12)}{1 - (1 + 0.13/12)^{-120}}$	0.13	0.13
	[÷]	[÷]
	12	12
	[=]	[=]
	[STO]	[STO]
	[+]	[+]
	1	1
	[y^x]	[=]
	120	[y^x]
	[+/−]	120
	[=]	[+/−]
	[+/−]	[=]
	[+]	[+/−]
	1	[+]
	[÷]	1
	15000	[=]
	[÷]	[÷]
	[RCL]	[RCL]
	[=]	[÷]
The calculation gives R = \$223.96611 or \$223.97. Over the life of the loan we pay (223.97)(120) = \$26 876.40.	[$1/x$]	15000
		[=]
		[$1/x$]

Example: Suppose we decide to pay off the loan with payments of R = \$111.98 every two weeks, that is, payments equal to half the monthly payment. How many periods will it take to pay off the loan and what will be the total amount paid over the life of the loan? Solving the equation above for the number of pay periods gives us

$$n = \frac{-\log(1 - A\,i/R)}{\log(1 + i)}$$

The interest rate each pay period might be calculated by using 365 days a year with payments every 14 days (many banks will, alternatively, use 360 days a year.)

	AL or AOS
Thus, $i = \dfrac{(0.13)}{(365/14)} = \dfrac{(0.13)(14)}{365}$ $n = \dfrac{-\log(1 - (15000)i/111.98)}{\log(1 + i)}$ The calculation gives $n = 221.63685$, that is, 221 payments of \$111.98 each with the last payment $(0.6368453)(111.98) = \$71.31$. The entire loan will be paid off in (222)(2) weeks, or approximately 8½ years. The total amount paid will be $(221)(111.98) + 71.31 = \$24\,818.89$.	0.13 [×] 14 [÷] 365 [=] [STO] [×] 15000 [÷] 111.98 [+/−] [+] 1 [=] [log] [+/−] [EXC] [+] 1 [=] [log] [1/x] [×] [RCL] [=]

2. *Arithmetic mean and standard deviation.* If $x_1, x_2, x_3, x_4, \ldots, x_n$ are a set of values (e.g., scores), the arithmetic mean is denoted by $\overline{x} = \dfrac{\Sigma x_i}{n}$. The amount of variation is commonly measured by the standard deviation, s:

$$s = \sqrt{\frac{\Sigma (x_i - \overline{x})^2}{n}} \quad \text{or} \quad s = \sqrt{\overline{x^2} - (\overline{x})^2}$$

(*Note:* The second formula for the standard deviation simply states that s is the square root of the difference between the mean of the squares and the square of the mean.) Addressable memory can be used to calculate the second formula above. As shown in chapter 3, we can compute the sum of the squares of the values in the display while concurrently computing the sum in the memory.

Example: Find the mean and standard deviation of the values 3, 7, 8, 1, 4.

	AL or AOS
	3
	[STO]
	[x^2]
	[+]
	7
	[M+]
	[x^2]
	[+]
	8
	[M+]
	[x^2]
	[+]
	1
	[M+]
	[x^2]
	[+]
	4
	[M+]
	[x^2]
Note: Sum of squares = 139 is in display.	[=]
	[÷]
	5
Note: n = 5 and the display is now $(\overline{x^2}) = 27.8$.	[=]
	[EXC]
	[÷]
	5
	[=]
Note: You may wish to record the mean = 4.6 in display.	[x^2]

	EXC
	−
	RCL
	=
Note: $s = 2.5768$ is in the display.	√

Many calculators are available with built-in present/future value or statistical functions, which cut down considerably on the number of keystrokes needed to perform these special calculations. Purchase of such a calculator should be considered by those who frequently need to make such calculations. Of course, a statistics text or business mathematics text should be consulted to understand the theory and conditions under which each of the formulas is appropriate.

Review

1. Is exponentiation, the y^x key, a binary operation or a unary function?
2. What is a pending operation?
3. Why should one be careful when using parentheses keys? What technique might be used to eliminate the need for them?
4. Suppose you need to calculate the expression $A - B$ (or $A \div B$), where A and B are expressions that must be evaluated before the subtraction (or division) can be performed. Describe an efficient algorithm for this type of computation.
5. Describe at least two practical situations for which a built-in constant might be a labor-saving feature on a calculator. Use two different operations for your examples.

Exercises

Section A.

Predict the result that would be given in the display of both AOS and AL calculators for the following keystroke sequences. Assume that both machines treat exponentiation as a binary operation that follows the rules applicable for that operating system.

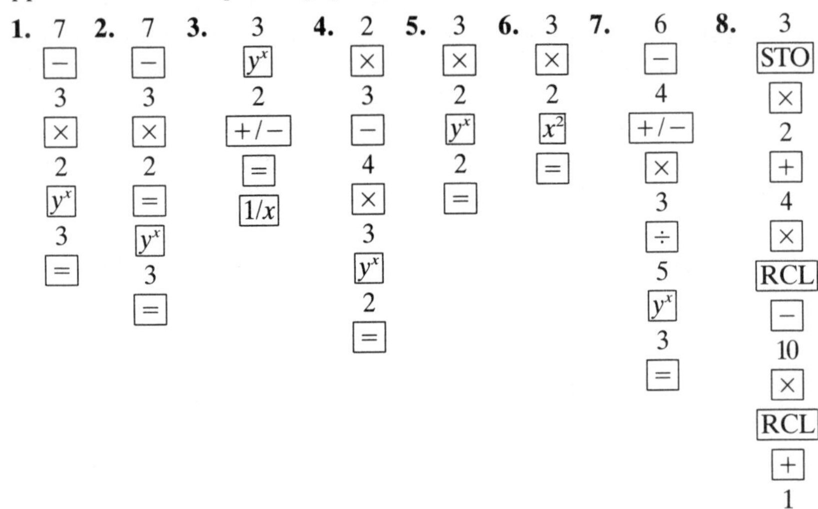

Section B.

Give AL and AOS keystroke sequences for calculating the following mathematical expressions.

9. $\sqrt[3]{5} = 5^{1/3}$ **10.** 5^{-3} **11.** $5^{-1/3}$ **12.** $2^{-0.25}$

13. $4(2) - 3^3$ **14.** $[4(2) - 3]^3$ **15.** $4(2^3) - 3^3$

16. $2 \times 5 - 4 \times 3^2$ **17.** $3^2 \times 5^2 - (4 + 7)^2$

18. $\dfrac{4 \times 7 - 3 \times 5}{4 \times 5 - 3 \times 7}$ **19.** $\dfrac{(11)(14)}{3 \times 5 - 3 \times 7 + 4 \times 7}$

20. $4 \times 7 - \dfrac{3 \times 5}{4 \times 5 - 3 \times 7}$ **21.** $5 + 3(6 - 5 \times 2^3)$

22. $(2 \times 7 - 3^2)^3 (3 \times 5 - 2 \times 4)^2 (3\sqrt{7} + 5)^{-2}$

23. $\dfrac{(2 \times 5 + 7)^3 - 4^3}{(2 \times 8 - 14 + 3)^3}$ **24.** $4 + 3 \times 5^{(37 - 5 \times 2^3)}$

25. Evaluate $[(3x - 4)x - 7]x - 2$ for $x = -2$

26. Evaluate $2 + x[7 - x(3x - 4)]$ for $x = -2$

27. Evaluate $\dfrac{5\,x^2 - 2x + 4}{3x^3 - x + 1}$ for $x = -1.238$

Section C.

Solve each of the following future-value and statistical problems using either an AL or an AOS calculator.

28. You can borrow $30 000 for 20 years at 14 percent interest, and plan to repay the loan in equal monthly payments. How much is each monthly payment? What is the total amount of money that you must repay?

29. In 28 above, suppose you elect to make payments every two weeks (14 days) that are equal to half of the monthly payments computed in 28. How many periods will be required to pay off the loan? How many years is this? What is the total amount of money repaid using this alternate repayment schedule?

30. Find the mean and standard deviation of the set of scores {3, 6, 7, 9, 2, 7}.

Section D.

The following keystroke sequences are entered and the indicated displays are observed in four different calculators. Fill in the final predicted display for each calculator.

Calculator	A	B	C	D
Keystroke sequence	3 [+] 5 [=]	3 [+] 5 [=]	3 [+] 5 [=]	3 [+] 5 [=]
DISPLAY	5	7	10	8
Keystroke sequence	2 [=]	2 [=]	2 [=]	2 [=]
DISPLAY	8	8	8	8
Keystroke sequence	4 [=]	4 [=]	4 [=]	4 [=]
PREDICTED DISPLAY				

Describe a type of application or problem in which each of the above calculators might be useful. Also indicate how you would operate the calculator to solve the problem.

5 Reverse Polish Notation (RPN)

REVERSE Polish Notation (RPN) is a calculator operating system that treats binary operations as real functions of two variables. Using the right-hand notation developed in chapter 2 on functions we have:

$$
\begin{aligned}
(2,4)+ &= 6\\
(2,4)- &= -2\\
(2,4)\times &= 8\\
(2,4)\div &= 1/2 = 0.5\\
(2,4)y^x &= 16
\end{aligned}
$$

The ordered pair (2,4) is acted on by the addition function to give 6, by the subtraction function to give -2, and so on. Analogous to functions of one variable in any operating system, functions of two variables do not require use of the [=] key. Thus, *there is no* [=] *key on the face of an RPN calculator.*

Since it is advantageous to disclose the contents of the operating registers in RPN, keystroke sequences for RPN will be represented differently from previously. The operating registers in RPN are represented by a stack. The first (lowest) level is sometimes called the x-register, the contents of which are shown in the display. The second level is the y-register, the third level the z-register, and the fourth level the w-register. All functions of a single variable are performed on the x-register. Functions of two variables (binary operations) are performed on the (y,x) ordered pair in the y- and x- registers. The small box displayed below and to the left of the stack gives the keystroke to be used.

When a number is keyed in, it enters the x-register. To perform a binary operation, it must be duplicated in the y-register. The ENTER ([ENT]) key copies the contents of the x-register into the y-register and moves the contents of each other register up one level. The next number keyed in replaces

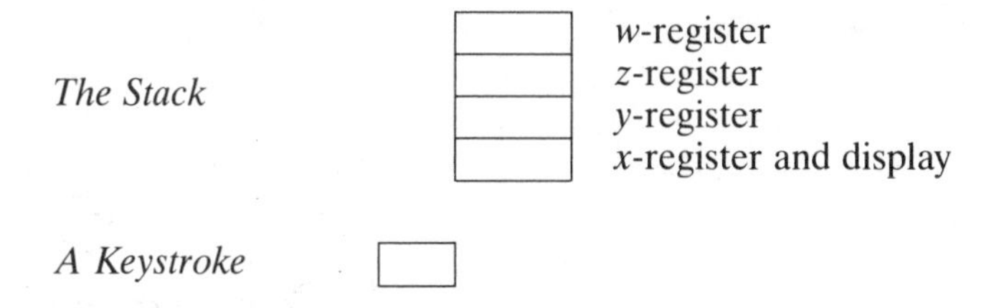

the contents of the x-register. When a binary operation is performed on the (y,x) ordered pair, the result appears in the display, and all values in the upper two registers move down one register. The following examples illustrate some simple binary operations:

Example 1

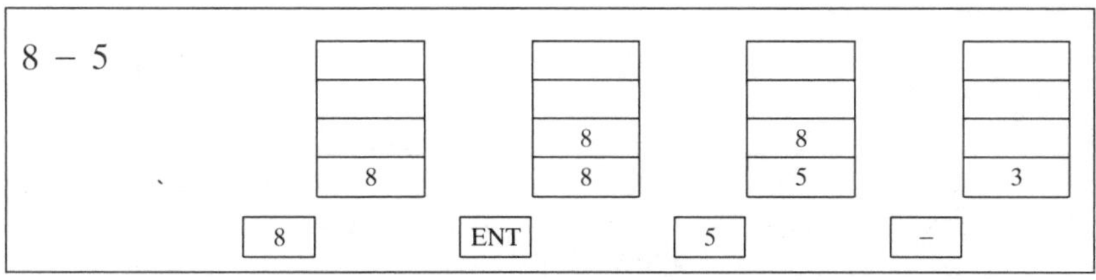

Without the ENT in the keystroke sequence above, the number 85 would have been entered into the x-register.

Example 2:

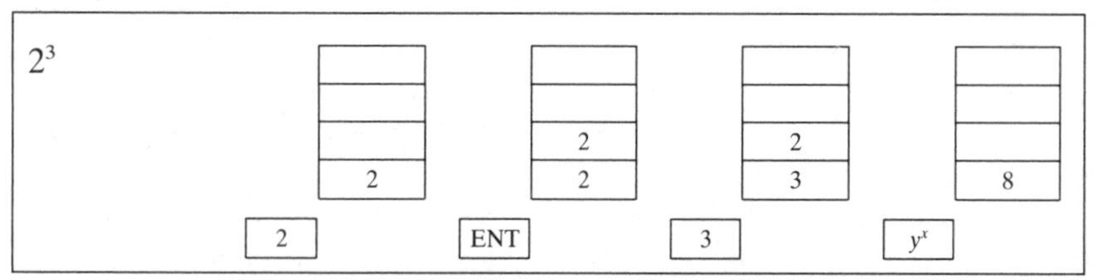

In example 3, after the first multiplication (before the 2 is keyed), observe that the ENTER key is *not* needed to move the contents of the x-register (15) up to the y-register. In general, any number keyed in will enter the x-register. The only time the ENTER key is necessary is when two distinct numerical values are keyed successively.

In this example, all multiplications are completed from left to right and then the subtraction is performed, similar to the way in which the computation would be done mentally. For many who are accustomed to performing mathematical calculations by hand, the left-to-right calculation of RPN is more natural than the algorithm in AL for which the subtrahend is first calculated and stored. Similarly, fractions are computed by first calculating the numerator, then the denominator, and finally the quotient, in contrast to the AL or AOS algorithms, which calculate from bottom to top.

Example 3

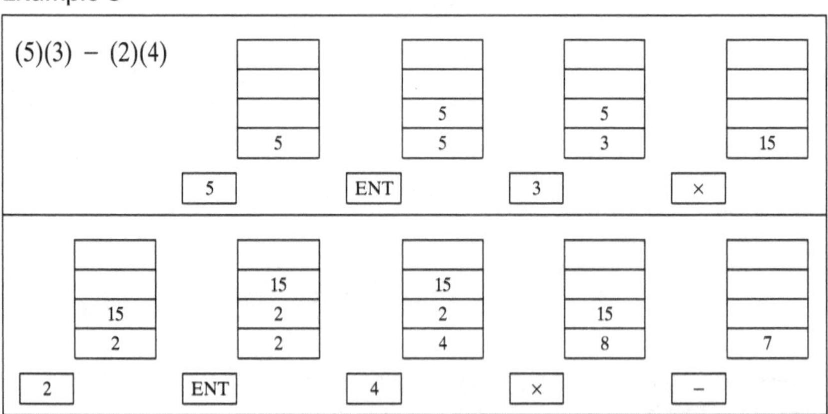

Example 4

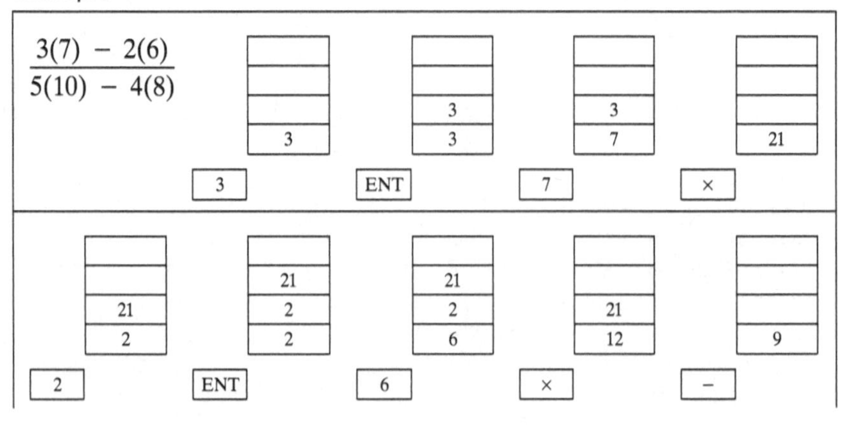

5	ENT	10	×	4
	9	9		9
9	5	5	9	50
5	5	10	50	4

ENT	8	×	−	÷
9	9			
50	50	9		
4	4	50	9	
4	8	32	18	0.5

All four levels of the stack were used in the solution of example 4. Some RPN calculators have only a three-level stack. If the keystroke sequence above were attempted on such a calculator, the number 9 would be lost, since there is no *w*-register, and the keystroke sequence above would probably yield the value zero (i.e., $0 \div 18 = 0$).

Functions of a single variable (e.g., $+/-$, x^2, $\sqrt{\ }$, $1/x$, sin, cos, log, etc.) are calculated with an RPN calculator in the same way as with any other calculator, namely, by pressing the desired function key.

Example 5

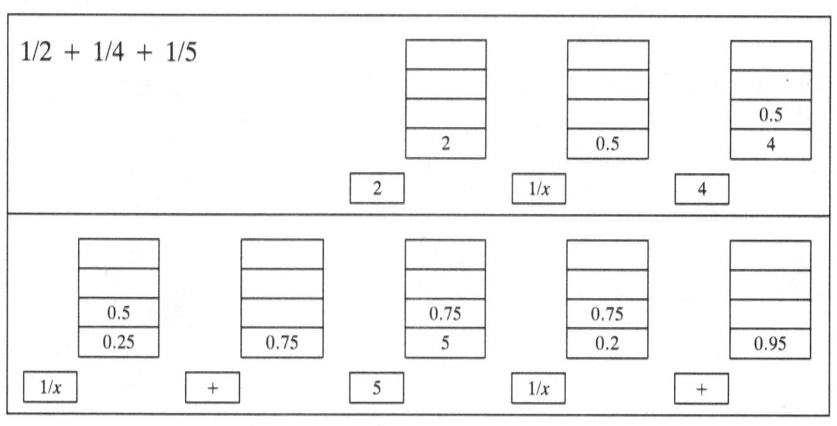

Many RPN calculators have one or more addressable memories that operate similarly to the memories on other calculators. The addressable memory should not be confused with the operating registers that make up the stack. The following present-value example (see also chapter 4, p. 44) shows the use of the memory to store a frequently used constant.

Example 6

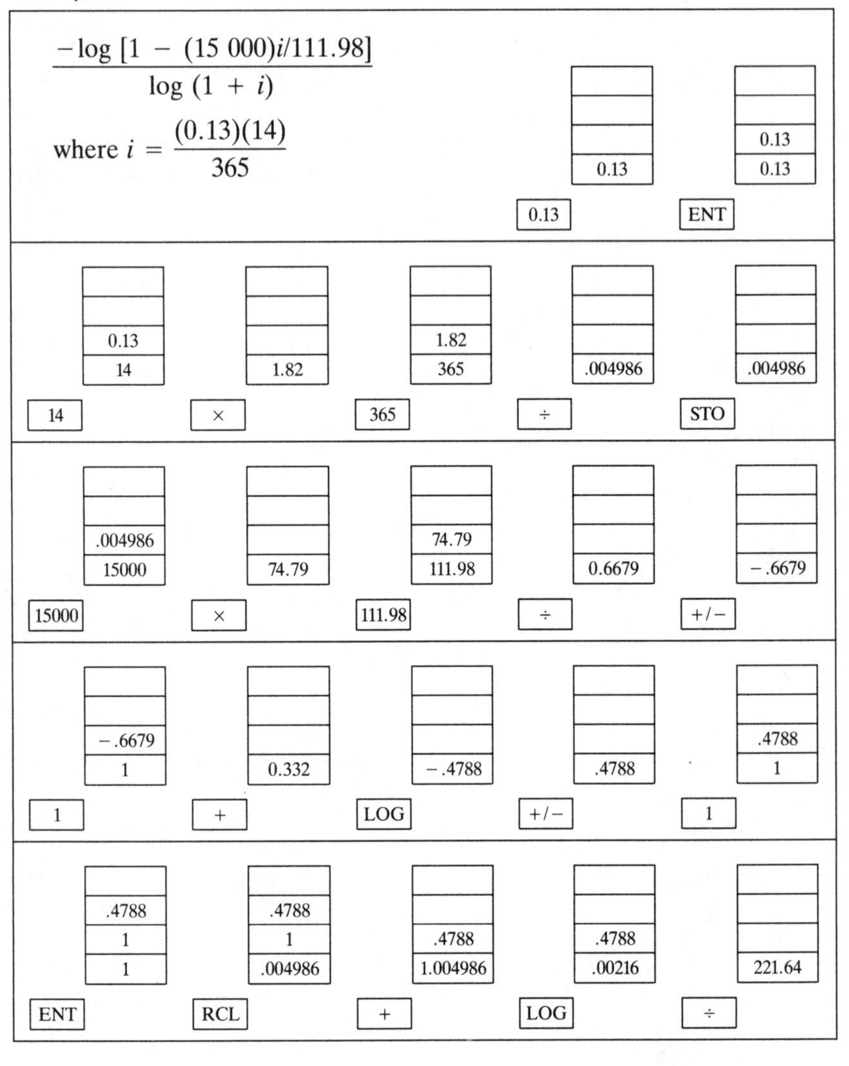

$$\frac{-\log[1-(15\,000)i/111.98]}{\log(1+i)}$$
where $i = \frac{(0.13)(14)}{365}$
0.13
0.13
0.13
0.13
ENT
0.13
14
1.82
1.82
365
.004986
.004986
14
×
365
÷
STO
.004986
15000
74.79
74.79
111.98
0.6679
−.6679
15000
×
111.98
÷
+/−
−.6679
1
0.332
−.4788
.4788
.4788
1
1
+
LOG
+/−
1
.4788
1
1
.4788
1
.004986
.4788
1.004986
.4788
.00216
221.64
ENT
RCL
+
LOG
÷

In general, the contents of the upper y-, z-, and w-registers are not seen by the calculator user, since only the contents of the x-register are shown in the display. However, many RPN calculators have a roll-down feature that transfers the contents of each upper register to the register below, while the contents of the x-register are transferred to the top w-register. Using the symbol R↓, the following example illustrates the use of the roll-down feature:

9	3
7	9
5	7
3	5
[ ]	R↓

Using the roll-down feature four times in succession allows the user to view the complete contents of the stack and returns the stack to its original configuration.

Review

1. Binary operations are treated as ______________________ in RPN.
2. What key is missing from an RPN calculator that is always found on an AL or AOS calculator?
3. In general if $*$ is any binary operation (e.g., $+$, $\times$, $-$, $\div$, y^x) describe the keystroke sequence and contents of the stack to calculate $a * b$, where a and b are any two real, positive numbers.
4. What does the ENTER ([ENT]) key do? When should it be used?
5. One way to compute 4^3 is to use the exponentiation key as a binary operation. Describe the necessary keystroke sequences and stack contents in order to use repeated multiplication as an alternate strategy to compute 4^3 in RPN.

Exercises

1–17. Write the keystroke sequences and show the contents of the RPN stack for the mathematical expressions listed in the exercises for chapter 4, pp. 48 and 49, numbers 9–27. Use a form similar to the one below to record your answers.

6 Working with Very Large or Very Small Numbers

CALCULATORS are known to be useful for particularly large or very small numbers. Depending on the calculator, attempting to calculate the expression

$$\frac{(6000)(8900)}{(0.00035)(0.014)}$$

may yield either no display, an error indication, or the display [1.0898 13.]. The last display is the result of the previous example written in scientific notation. Traditionally, the display above would be written 1.0898×10^{13}.

For those who are not familiar with scientific notation, a short review is presented. A number written in scientific notation is expressed as a decimal between 1 and 10 (i.e., having one nonzero digit to the left of the decimal point) multiplied by a power of ten.

$$\begin{aligned}
100000 &= 10^5 \\
10000 &= 10^4 \\
1000 &= 10^3 \\
100 &= 10^2 \\
10 &= 10^1 \\
1 &= 10^0 \\
0.1 &= 10^{-1} = 1/10 \\
0.01 &= 10^{-2} = 1/100 \\
0.001 &= 10^{-3} = 1/1000 \\
0.0001 &= 10^{-4} = 1/10000
\end{aligned}$$

Numbers larger than 1 have positive exponents, whereas those smaller than 1 have negative exponents.

To change 3689 to scientific notation we obviously need to move the decimal three places to obtain a coefficient with one digit to the left of the decimal, and since 3689 is larger than 1, the exponent will be positive. Hence,

$$3689 = 3.689 \times 10^3.$$

To change 0.0257 to scientific notation, we must move the decimal point two places to the right, and since 0.0257 is smaller than 1, the exponent will be negative. Hence,

$$0.0257 = 2.57 \times 10^{-2}.$$

To reverse the process, 3.689×10^3 obviously represents a number larger than 3.689 since the exponent is positive. Move the decimal point three places in a direction to reflect that 3.689×10^3 is a number larger than 3.689 (i.e., $3.689 \times 10^3 = 3689$). The expression 2.57×10^{-2} represents a number smaller than 2.57; to convert to standard notation, move the decimal point two places in a direction that reflects that 2.57×10^{-2} is a number smaller than 2.57 (i.e., $2.57 \times 10^{-2} = 0.0257$).

Numbers expressed in scientific notation are easily manipulated using the standard laws of exponents.

General law of exponents	Applied to scientific notation (i.e., $b = 10$)
$b^m \cdot b^n = b^{m+n}$	$10^m \cdot 10^n = 10^{m+n}$
$\frac{1}{b^n} = b^{-n}$	$\frac{1}{10^n} = 10^{-n}$
$(b^n)^m = b^{nm}$	$(10^n)^m = 10^{nm}$
$(ab)^n = a^n b^n$	$(k \times 10^m)^n = k^n \times 10^{mn}$

No matter what type calculator you use, by applying the above laws of exponents it is easy to calculate expressions involving very large or very small numbers. First convert all the numbers in the expression to scientific notation. Then operate on the coefficients using the calculator, while computing the exponents using the laws of exponents above. For calculators with scientific notation, this alternative strategy has the advantage of increased precison over that obtainable from using the scientific notation facility. For calculators without scientific notation, there is no other way to compute with very large or very small numbers.

Example 1

$$\begin{aligned}
\frac{(6000)(8900)}{(0.00035)(0.014)} &= \frac{(6 \times 10^3)(8.9 \times 10^3)}{(3.5 \times 10^{-4})(1.4 \times 10^{-2})} \\
&= \frac{6 \times 8.9}{3.5 \times 1.4} \times 10^3 \times 10^3 \times 10^4 \times 10^2 \\
&= 10.897959 \times 10^{12} \\
&= 1.0897959 \times 10^1 \times 10^{12} \\
&= 1.0897959 \times 10^{13}
\end{aligned}$$

For illustrative purposes we have included all the digits that an eight-digit display calculator might show. For most purposes such precision is not warranted.

Example 2

$$\begin{aligned}
(4 \times 10^{-3})^2 (6 \times 10^5)^{-3} &= 4^2 \times 10^{-6} \times 6^{-3} \times 10^{-15} \\
&= \frac{4^2}{6^3} \times 10^{-6} \times 10^{-15} \\
&= \frac{16}{216} \times 10^{-21} \\
&= 0.0740741 \times 10^{-21} \\
&= 7.40741 \times 10^{-2} \times 10^{-21} \\
&= 7.40741 \times 10^{-23}
\end{aligned}$$

If greater precision is desired, the expression in step 3 could be rewritten as

$$\begin{aligned}
\frac{16 \times 10^{-21}}{216} &= \frac{16 \times 10^{-21}}{2.16 \times 10^2} \\
&= \frac{16}{2.16} \times 10^{-21} \times 10^{-2} \\
&= 7.4074074 \times 10^{-23}.
\end{aligned}$$

Example 3

$$
\begin{aligned}
(3 \times 10^{142}) - (5 \times 10^{144}) &= (3 \times 10^{142}) - (5 \times 10^{142} \times 10^{2}) \\
&= (3 - 5 \times 10^{2}) \times 10^{142} \\
&= (3 - 500) \times 10^{142} \\
&= -497 \times 10^{142} \\
&= -4.97 \times 10^{2} \times 10^{142} \\
&= -4.97 \times 10^{144}
\end{aligned}
$$

Here the distributive law was used to factor a common factor of 10^{142} from both terms of the sum. An alternative strategy would be to factor 10^{144} from both terms.

$$
\begin{aligned}
(3 \times 10^{142}) - (5 \times 10^{144}) &= (3 \times 10^{-2} - 5) \times 10^{144} \\
&= (0.03 - 5) \times 10^{144} \\
&= -4.97 \times 10^{144}
\end{aligned}
$$

The number 10^{144} is a very large number! Seldom would such a large number be used in any calculation. Even calculators with scientific notation usually only have exponents ranging from -99 to $+99$. However, as is seen above, working with exponents outside of this range is not difficult.

Calculators having scientific notation will automatically change answers that exceed display limits into scientific notation. Numbers may also be entered directly in scientific notation into such calculators. Typically there is a key marked [EE] or [EX] to indicate that the number about to be entered is the exponential part of such a number. The following keystroke sequences illustrate how to enter a number in scientific notation:

Number	3.02×10^{8}	-3.02×10^{8}	3.02×10^{-8}	-3.02×10^{-8}
Keystroke sequence	3.02 [EE] 8	3.02 [+/−] [EE] 8	3.02 [EE] 8 [+/−]	3.02 [+/−] [EE] 8 [+/−]

In each case, the coefficient—including the sign of the coefficient—is entered first. On entering the key [EE] (or other key to designate exponent), a space and the two digits 00 appear in the display. The exponent and the sign of the exponent are then entered. The keystrokes for computing example 2, page 59, using the scientific notation capability of the calculator are as follows:

Mathematical expression	*AL or AOS*	*RPN*
$(6 \times 10^5)^{-3} (4 \times 10^{-3})^2$	6 [EE] 5 [y^x] 3 [+/−] [×] 4 [EE] 3 [+/−] [x^2] [=]	6 [EE] 5 [ENT] 3 [+/−] [y^x] 4 [EE] 3 [+/−] [x^2] [×]

For most of the discussion thus far, complete eight-digit calculator display results have been shown. Such precision is not warranted in most applications. The general rule of thumb is that the final answer should have the same number of significant digits as the least precise number in the computation. When numbers are written in scientific notation, the number of significant digits is always evident. It is difficult to tell how many significant digits are in a number written as 7000. However, written as 7.0×10^3, it has two significant digits, whereas written as 7.000×10^3, it has four significant digits. If a number is to be used in further calculations, it is important to retain at least one or two additional significant digits beyond the number wanted in the final answer. If operations are chained, and intermediary answers retained electronically with the calculator, the full eight- or ten-digit precision is used for intermediary results.

When nearly all the available display digits are used, one must be concerned about machine round-off error. Calculators are finite machines that operate with finite, truncated decimals, not infinite decimals. On a calculator with an eight-digit display, try the following calculation, working from left to right: 0.0000008 + 1234567 − 1234567. (On a ten-digit display, enter the number 0.000000008 and proceed with the rest of the problem as written above; the idea is that we want a number less than 1 that fills up the display.) Each of the numbers in the expression will appear on a eight-digit display, but the intermediate answer 1234567.0000008 cannot be displayed exactly.

A bit of experimentation with this example can lead to important discoveries about your own calculator. Try each of the following problems:

$$0.0000008 + 12 - 12$$
$$0.0000008 + 123 - 123$$
$$0.0000008 + 1234 - 1234$$
$$0.0000008 + 12345 - 12345$$

Both the intervening and the final displays will give insights into the calculator you are using. For example, one of my calculators with an eight-digit display gives the following results:

Example	*A*	*B*	*C*
Keystroke sequence	.0000008 [+] 1234 [−] 1234 [=]	.0000008 [+] 12345 [−] 12345 [=]	.0000008 [+] 123456 [−] 123456 [=]
DISPLAY	0.0000008	0.000001	0.

D	*E*
.0000008 [+] 12 [=]	.0000008 [+] 12 [−] 12 [=]
12.000001	0.0000008

We could infer that in example A, the intermediate answer 1234.0000008 is being retained internally in the calculator; in example B, the intermediate answer 12345.0000008 has been rounded to 12345.000001 in its internal (electronic) representation; and in example C, the intermediate answer 123456.0000008 has been truncated to 123456, or since machines operate consistently, rounded to 123456.00000. We would conclude that even though there are eight digits in the display, this calculator probably carries eleven digits in its operating (internal) registers. Example D indicates that this calculator also rounds off displayed results, even though we believe (and verify in example E) that the value 12.0000008 is retained internally.

The existence of additional internal precision for a machine may be another argument for chaining operations when using a calculator rather than writing down intermediate results and rekeying them. (See chapter 3, p. 28, for the argument previously given.) If intermediate results are written down or if your calculator does not carry additional digits internally then the following examples will produce results inconsistent with the usual distributive, associative, and commutative laws of arithmetic.

Example 4

Assuming intermediate eight-digit display results are rekeyed or the calculator does not carry additional digits internally, the distributive law may appear to be violated.

.5(5.0000004 + 9.0000004)	vs.	.5(5.0000004) + .5(9.0000004)
Rounded eight-digit results	Truncated eight-digit results	Either rounded or truncated eight-digit results
.5(14.000001) 7.0000005	.5(14) 7.	2.50000002 + 4.50000002 7.0000004

As shown, adding the 5.0000004 and 9.0000004 on the left makes the display overflow. If the answer is either truncated or rounded and then used to multiply by .5, the result will differ from that on the right.

Example 5

Assuming intermediate eight-digit display results are rekeyed or the calculator does not carry additional digits internally, the commutative (or associative) laws may appear to be violated.

Eight-digit results	12.000001 + 1.0000003 + 3.0000003 + 2.0000004 = 13.000001 + 3.0000003 + 2.0000004 = 16.000001 + 2.0000004 = 18.000001
	vs.
Eight-digit results	2.0000004 + 3.0000003 + 1.0000003 + 12.000001 = 5.0000007 + 1.0000003 + 12.000001 = 6.0000010 + 12.000001 = 18.000002

In the top computation, each succeeding addition exceeds the eight-digit display and either truncation or rounding will result in losing the trailing digit (last digit on right in exact answer), giving a final answer of 18.000001 (if additional digits are not carried internally). On the bottom row, no trailing digits are lost in the first three additions. The sum of the first three trailing digits carries into the next place value to the left and the final result of the addition on the bottom row is 18.000002.

Clearly, the calculator number system consists of a finite number of decimal fractions. Further, the points of the calculator are not equally spaced. On an eight-digit display calculator, the numbers in the range $(-10, +10)$ are equally spaced with a spacing of 0.0000001. For example, the successive numbers 1.0000003 and 1.0000004 differ by 10^{-7}. In the range $[+10, +100)$ and $(-100, -10]$, values are equally spaced with a spacing of 0.000001. For example, the successive numbers 10.000003 and 10.000004 differ by 10^{-6}. At most, eight-digit numbers *can be entered* into the calculator or displayed even though the number of digits retained internally may be larger than eight. The internal representation is also of a finite number of digits unevenly spaced even though this spacing may be different from the spacing on the face of the calculator. Most of the time these differences from our conceptual idea of a continuous number system cause few difficulties. However, when many calculations are involved, one must be aware of the possibilities of machine round-off error.

Review

1. Give a simple example to illustrate machine round-off error.
2. 10^a is (less than) or (greater than) 1 when a is ______________________.
3. Numbers represented in scientific notation are represented as a product of a power of ten and ______________________.

Exercises

Section A.

Write each of the following numbers in scientific notation:

1. 304 **2.** 80 000 000 **3.** 0.3 **4.** 0.0056 **5.** 6.49

Section B.

Write each of the following numbers in decimal form:

6. 10^3 **7.** 10^{-4} **8.** 10^{-1} **9.** 10^0

10. 3.12×10^{-3} **11.** 4.05×10^4 **12.** 9.00×10^{-2}

13. 2.300×10^5 **14.** 3.723×10^0

Section C.

Complete each of the following:

15. $0.013 \times 10^{-3} = 1.3 \times$ ________

16. $295.4 \times 10^{2} = 2.954 \times$ ________

17. $3.75 \times 10^{3} =$ ________ $\times 10^{-1}$

18. $6.03 \times 10^{-2} =$ ________ $\times 10^{-1}$

19. $0.6 \times 10^{-2} =$ ________ $\times 10^{-4}$

20. $63.75 \times 10^{6} =$ ________ $\times 10^{3}$

Section D.

Perform the indicated operations:

21. $\dfrac{(3 \times 10^{7})(3 \times 10^{-4})}{6 \times 10^{3}}$

22. $\dfrac{(16 \times 10^{-3})(3 \times 10^{4})}{1.2 \times 10^{-6}}$

23. $\dfrac{(3.5 \times 10^{4})(12 \times 10^{-1})}{2.1 \times 10^{-8}}$

24. $(6 \times 10^{-25})^{2}$

25. $(2 \times 10^{19})^{3}$

26. $3.07 \times 10^{4} + 3.9 \times 10^{4}$

27. $3 \times 10^{2} + 5 \times 10^{-2}$

28. $6 \times 10^{23} - 7 \times 10^{21}$

29. $4.5 \times 10^{151} - 7.8 \times 10^{154}$

30. $5.4 \times 10^{-141} + 3.7 \times 10^{-144}$

31. $(6.0 \times 10^{5})^{2}(3 \times 10^{4})^{-3}$

32. $(3 \times 10^{-4})^{2}(5 \times 10^{7})^{-3}$

33. $\dfrac{(237\ 000)(0.000374)(305)}{(0.091)(5100)(100.5)}$

34. $\dfrac{(589\ 060\ 000)(639)(87\ 249)}{(0.00158)(0.0034)(0.05906)}$

7
Approximating Solutions to General Polynomials and Transcendental Equations

ALTHOUGH elementary algebra techniques can be used to solve linear equations (e.g., $3x + 5 = 0$) or quadratic equations (e.g., $x^2 + x - 2 = 0$), they are not suitable for solving the general polynomial or transcendental equation. In fact, for the general polynomial of degree 5 or greater, it can be shown that there is no general algebraic solution. However, we can estimate the solution to any equation of the form $f(x) = 0$ by graphing. The graph of $y = f(x)$ crosses the x-axis at the points where $y = 0$, hence the solution(s) for $f(x) = 0$ are those points x at which the graph of $y = f(x)$ crosses the x-axis.

Example: Find the solutions to $x^3 - 4x + 1 = 0$. Plotting $y = x^3 - 4x + 1$ on the plane, we obtain the graph below that crosses the x-axis in three

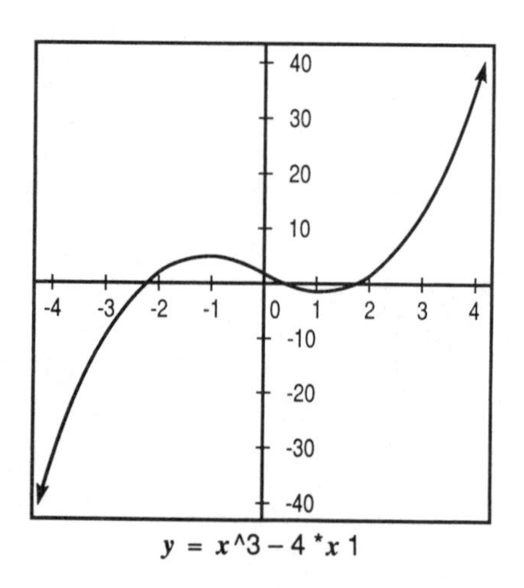

$y = x$^3 – 4 *x 1

places, showing the equation has three roots. The smallest root satisfies $-3 < x < -2$. The next root satisfies $0 < x < 1$. The largest root satisfies $1.5 < x < 2.5$. The graph has allowed us to estimate the solutions to the nearest whole number, namely $x = -2$, $x = 0$, and $x = 2$.

Suppose we needed greater precision for the middle root, 0. Since we know that $0 < x < 1$, we might plot an expanded graph for the domain [0, 1], as shown below. An examination of this graph reveals that $0.2 < x < 0.3$, and allows us to estimate x to the nearest tenth of a unit, $x = 0.25 \pm 0.05$. Continuing the same process, the next graph shows that $0.25 < x < 0.26$ and the graph on the right shows that $0.254 < x < 0.255$.

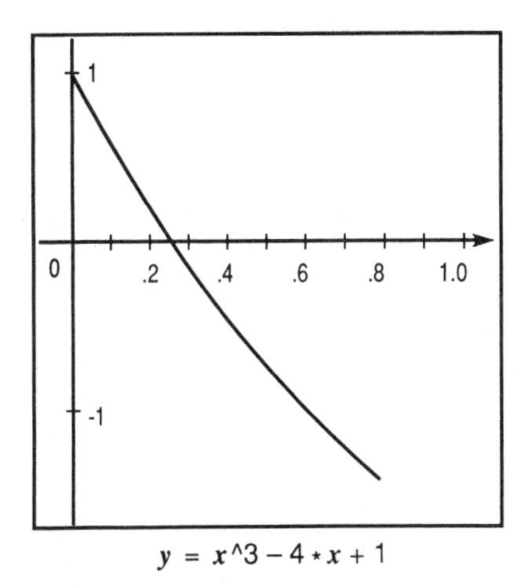

y = x^3 – 4 * x + 1

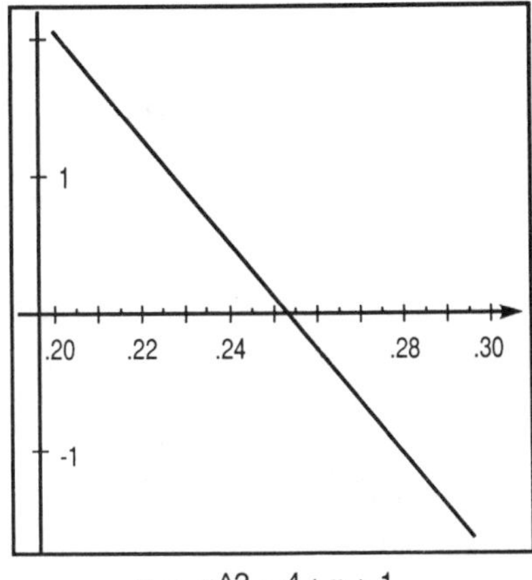

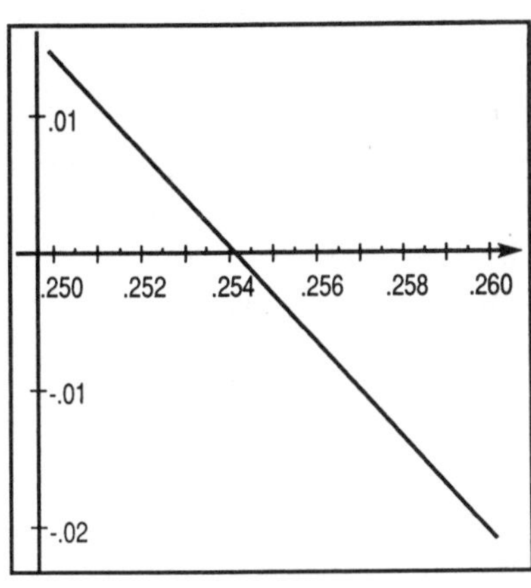

y = x^3 – 4 * x + 1

Hence, we would conclude that to three decimal places, the middle root = 0.254. Obviously the same process can be continued to achieve whatever precision is required.

In the section below we will discuss how to use the calculator efficiently to obtain the functional values needed to plot such graphs and will discover that, in practice, we will need far fewer functional values to estimate a solution to the required degree of precision than would be required to accurately draw the graphs.

If one attempts to evaluate $(-4)^3$ directly on the calculator, many machines will produce an error condition, even though $(-4)^3 = -64$ is a well-defined real number. This unfortunate operating characteristic makes direct calculation of polynomial functions (e.g., $x^3 - 4x + 1$) awkward for negative x-values.

Nested interval decomposition provides an efficient algorithm that can be used to calculate ordered pairs that satisfy any polynomial. The following examples illustrate how to perform the algebraic manipulations:

Example 1

$$\begin{aligned} x^3 - 4x + 1 &= [x^2 - 4]x + 1 \\ &= [(x)(x) - 4]x + 1 \end{aligned}$$

Example 2

$$\begin{aligned} &3x^4 - 5x^3 + 4x^2 + 7x - 8 \\ &= \{3x^3 - 5x^2 + 4x + 7\}x - 8 \\ &= \{[3x^2 - 5x + 4]x + 7\}x - 8 \\ &= \{[(3x - 5)x + 4]x + 7\}x - 8 \end{aligned}$$

In each successive step, the largest exponent is decreased by 1 using the right distributive property applied to all terms still having a power of x. As illustrated above, the process can be continued until all exponents are 1. Of course, for machines with a squaring function, exponents of 2 are quite manageable.

The AL, AOS, and RPN keystroke sequences for calculating example 1, using nested interval decomposition, are illustrated below for the case of $x = -4$. The AOS user should be careful to remember to use the [=] key to close parentheses.

Mathematical expression	*AL*	*AOS*	*RPN*
$f(x) = x^3 - 4x + 1 =$ $(x^2 - 4)x + 1$	4	4	4
	[+/−]	[+/−]	[+/−]
	[STO]	[STO]	[STO]
	[x^2]	[x^2]	[x^2]
	[−]	[−]	4
	4	4	[−]
	[×]	[=]	[RCL]
	[RCL]	[×]	[×]

	[+] 1 [=]	[RCL] [+] 1 [=]	1 [+]

When other values are substituted for x (and stored in the first steps above), the following table of (x, y) pairs satisfying the equation $y = x^3 - 4x + 1$ can be similarly obtained.

x	-4	-3	-2	-1	0	1	2	3	4
y	-47	-14	1	4	1	-2	1	16	49

Note that as x increases from -3 to -2, y increases from -14 to 1. Since $f(x)$ is a continuous function, all values between -14 and 1 must be achieved for some value of x between -3 and -2. Hence, there must be a root of the equation $f(x) = x^3 - 4x + 1 = 0$ in the interval $(-3, -2)$ or $-3 < x < 2$. We can now use a guess-and-check method to solve the equation. We are looking for a value of x that makes y close to 0. Noting that $f(-2) = 1$ is closer to 0 than $f(-3) = -14$, we might first guess that $x = -2.2$. Using the algorithm above, we can compute $f(-2.2) = -0.848$, giving us $-2.2 < x < -2.0$. Trying $x = -2.1$, we have $f(-2.1) = 0.139$. Since the point $(-2.1, 0.139)$ is above the x-axis, whereas the point $(-2.2, -0.848)$ is below it, we know the desired solution is in the interval $-2.2 < x < -2.1$. In general, solutions to the equation $f(x) = 0$ are always between x-values whose corresponding y-values have opposite signs.

Carrying the approximation process just described a step further, we can show that the desired root lies between -2.12 and -2.11, when calculated to the nearest hundredth. The process of calculation can be illustrated conveniently in the following table. Positive values of y are above the x-axis, of course, and negative values are below it. Of course, the other two roots, $x = 0.25$ and $x = 1.86$, can be similarly estimated.

x	y	Above or below x-axis
-3	-14	Below
-2	1	Above
-2.2	-0.848	Below
-2.10	$+0.139$	Above
-2.14	-0.24	Below
-2.12	-0.048	Below
-2.11	$+0.046$	Above

A similar guess-and-check method might be used to solve many transcendental equations.

Example 3

Solve $\cos x = x$.

The form of the equation makes it clear that x is a real number and hence, $\cos x$ must be evaluated for *x expressed in radians*. Graphing $y = \cos x$ and $y = x$ on the same axis we obtain the graph below. The two graphs intersect for $0 < x < 1$, therefore, the equation $\cos x = x$ has a solution in the interval $(0, 1)$.

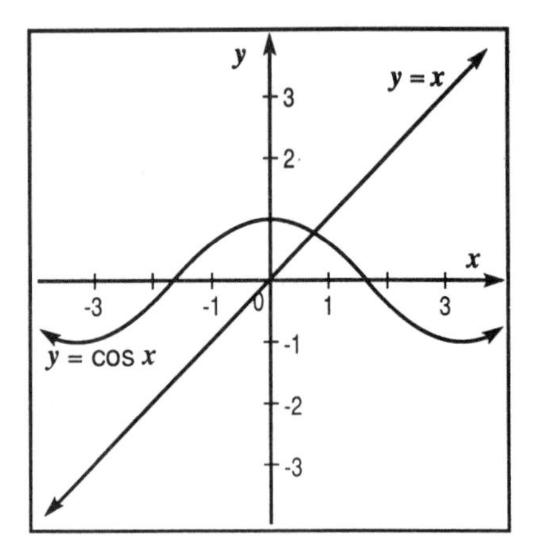

We also note that solving the equation $\cos x = x$ is equivalent to solving $x - \cos x = 0$. Thus we can use the same procedure we used in the previous example to find the roots of $y = f(x) = x - \cos x$, being sure to have the calculator in radian mode. From the table below, we find a solution in the interval $(0.73, 0.74)$. We could, of course, obtain greater precision by continuing the strategy:

x	y
0	−1
1	+0.460
0.7	−0.0648
0.8	+0.103
0.75	+0.018
0.73	−0.015
0.74	+0.0015

Although the guess-and-check method of solution is applicable to a wide variety of problems, the calculations involved may become impossibly laborious without some sort of calculating machinery.

Example 4

An open rectangular box is to be constructed from a rectangular piece of cardboard 16 inches wide and 23 inches long by cutting out a square from each corner and then bending up the sides. Find the size of the corner square that will produce a box having the largest possible volume.

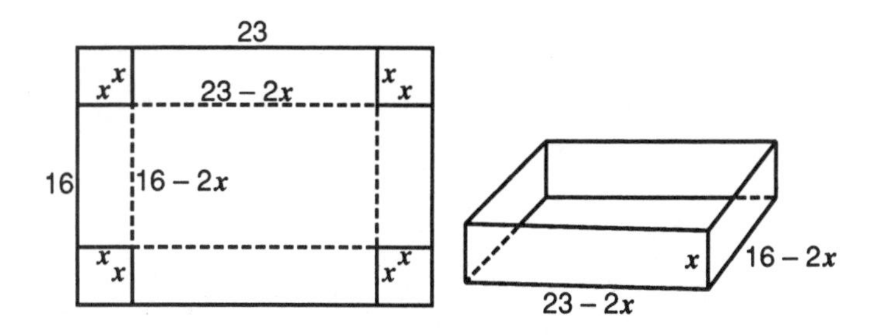

Letting *l*, *w*, and *h* stand for the length, width, and height, respectively, of the completed box, we have from the diagrams above: $l = 23 - 2x$; $w = 16 - 2x$; and $h = x$.

Hence, the volume,

$$\begin{aligned} V = lwh &= (23 - 2x)(16 - 2x)(x) \\ &= 4x^3 - 78x^2 + 368x \\ &= [4x^2 - 78x + 368]x \\ &= [(4x - 78)x + 368]x. \end{aligned}$$

Since, of course, all dimensions of the box must be positive, we have further that

$$h = x > 0$$

and

$$w = 16 - 2x > 0, \text{ or } x < 8.$$

This allows us to restrict our search to the interval $0 < x < 8$. We can proceed as illustrated in the table below to find the value of x in this interval for which V is largest. A look at the even numbers in the interval shows us that we should search further in the interval (2,6). If we add in the odd integers in the interval, the value of $V = 510$ at $x = 3$ suggests further exploration of the interval (2,4). In the next step (units of 0.5), we further restrict the search to (2.5, 3.5). Going down to units of .1, we find a maximum value of V (510.384) at $x = 3.1$. *Note:* With algebraic techniques, the guess-and-check method can be used to estimate to any necessary degree of precision the value of x that gives a maximal value of V. If one uses calculus techniques, the problem can be reduced to $12x^2 - 156x + 368 = 0$. This quadratic equation, of course, can be solved using the quadratic formula, obtaining $x = 3.09657$ and $V = 510.3845$.

x	V
0	0
2	456
4	480
6	264
8	0
3	510
5	390
2.5	495
3.5	504
2.8	506.688
3.2	509.952
2.9	508.776
3.1	510.384

Review

1. Explain the relationship between the graph of $y = f(x)$ and the solution(s) to the equation $f(x) = 0$.
2. Using the guess-and-check method, we can estimate the solution to $f(x) = 0$ to any necessary degree of precision. Why is this sufficient for most physical problems?
3. List advantages and disadvantages of the guess-and-check method of solution.

Exercises

Section A.

For each of the polynomials, $f(x)$, below—

a) Rewrite the polynomial in nested interval decomposition;

b) Make a table of x and $y = f(x)$ values that describe the function throughout the plane;

c) Graph $y = f(x)$ from the points listed from the table in part (b);

d) Determine how many solutions the equation $f(x) = 0$ has;

e) Give integral bounds for each solution, that is, find integers a and $a+1$ such that $a < x < a+1$;

f) Estimate each solution of $f(x) = 0$ to the nearest 0.01 by using a guess-and-check technique.

1. $f(x) = x^3 - 3x^2 - 9x + 30$
2. $f(x) = 0.3x^3 - 1.8x^2 + 1.5x + 5$
3. $f(x) = 5x^3 - 26x^2 - 29x + 100$
4. $f(x) = x^4 - 5x^3 + 1$
5. $f(x) = 4x^3 - 19x^2 - 25x + 80$
6. $f(x) = 2x^3 + 7x^2 + 10$
7. $f(x) = 4x^2 - 2x^4 - 1$

Section B.

Solve each of the following equations by estimating x to the nearest 0.01:

1. $1/(x^2 + 1) = 2x$
2. $2 \sin x = x$, for $x > 0$
3. $\sin x = x^2$
4. $\tan x = x$, for $1.57 < x < 4.71$
5. $2x = 5 + \sin x$
6. $2x = e^{-x}$
7. $2x^3 + e^x = 0$

Section C.

Use the guess-and-check method to solve each of the following problems:

1. An open rectangular box is to be constructed from a rectangular piece of cardboard 36 cm wide and 50 cm long by cutting out a square from each corner and then bending up the sides. Find the size of the corner square that will produce a box having the largest possible volume.

2. A fence 7 feet tall stands on level ground and runs parallel to a tall building. If the fence is 1 foot from the building, find the shortest ladder that will extend from the ground over the fence to the wall of the building.

Hints:

a) Draw a picture of the fence and building as viewed from the end of the fence. Put the ladder in the picture.

b) Label all known lengths. Let x be the distance from the fence to the foot of the ladder. Let s be the distance from the top of the ladder to the ground.

c) Use similar triangles to express the distance s in terms of x.

d) Use the Pythagorean theorem to find the length of the ladder. [You have now expressed the length of the ladder, L, as a function of x.] Simplify the expression to allow for easy calculator computation of functional values.

e) Fill in the following table:

x	1	2	4	6	8
L					

f) At what value x, was the length of the ladder smallest?

g) Is this the shortest ladder possible?

h) If you were to look for a value of x which produced a shorter ladder, which values would you pick? ______ $< x <$ ______

i) Using this suggested domain for x, find the value of x (to the nearest 0.1) that minimizes L.

3. Find the radius and height of the (right-circular) cylinder of largest volume that can be inscribed in a (right-circular) cone with radius 5 inches and height 14 inches.

Hints:

a) Diagrams of the cylinder inscribed in a cone and a cross-sectional view of half the cone are shown below:

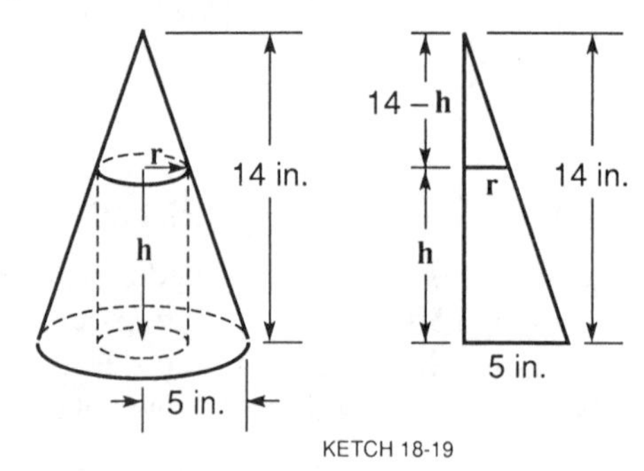

b) Using similar triangles from the diagram above, write an equation relating h and r.

c) Solve the equation from part (b) for h.

d) The formula for the volume of a cylinder is $V = \pi r^2 h$. Substitute the value of h found in part (c) into the expression for V, obtaining V as a function of r.

e) By using guess-and-check methods, find the value of r (to the nearest 0.1) that gives a maximal value for V.

f) Substitute the value of r found in part (e) into the expression for h found in part (c) to obtain the complete dimensions of the cylinder.

4. Assume that the operating cost of a truck (excluding driver's wage) is $12 + x/5$ cents per mile when the truck travels at a rate of x miles per hour. If the driver earns $6 per hour, what is the most economical speed to operate the truck?

Hints:

a) For any trip the distance is a constant. Without loss of generality, we can assume the distance traveled to be 1. This assumption is equivalent to finding the cost per mile traveled. (If this strategy is not clear, first assume the distance to be 1, work the problem, and then go back and rework the problem with distance = 100.)

b) How much does it cost to run the truck 1 mile?

c) Using $d = rt$, and knowing $r = x$ (given) and $d = 1$ (part (a)), we can find the time the driver takes in terms of x.

d) What is the driver cost?

e) What are the units for the cost of the truck in part (b)? What are the units for the cost of the driver in part (d)? Adjust so that units are the same (i.e., both cents or both dollars).

f) Express total cost, C, as a function of x.

g) Using guess-and-check methods and your own intuition of reasonable values of x, find the minimal cost C for x to the nearest whole number.

Answers to Exercises

Chapter 1

Section A

1. AL 16
AOS 11

2. AL 0.5
AOS 8

3. AL 12
AOS 12

4. AL 22
AOS 7

5. AL 13
AOS 23.5

6. AL 12
AOS 14

7. AL 12
AOS 12

8. AL 3
AOS 3

9. AL 70
AOS 50

10. AL 2
AOS −0.5

11. AL 1.25
AOS 2

12. AL 14
AOS −0.5

Section B

13.

AL	AOS
3	5
[×]	[+]
2	3
[+]	[×]
5	2
[=]	[=]

14.

AL	AOS
5	5
[+]	[+]
3	3
[×]	[=]
2	[×]
[=]	2
	[=]

15.

AL	AOS
2	2
[×]	[×]
3	3
[÷]	[−]
4	5
[−]	[×]
5	4
[×]	[=]
4	
[=]	

16.

AL	AOS
3	3
[÷]	[−]
4	7
[−]	[×]
7	4
[×]	[=]
4	
[=]	

17.

AL	AOS
2	2
[÷]	[÷]
3	3
[×]	[+]
7	4
[+]	[÷]
4	7
[÷]	[=]
7	
[=]	

18.

AL	AOS
3	3
[×]	[×]
7	7
[−]	[−]
9	9
[÷]	[=]
2	[÷]
[÷]	2
6	[÷]
[=]	6
	[=]

19.

AL	AOS
2	2
[×]	[×]
3	3
[÷]	[−]
5	4
[−]	[×]
4	5
[×]	[+]
5	7
[÷]	[×]
3	3
[+]	[=]
7	
[×]	
3	
[=]	

20. AL or AOS

21
[÷]
7
[×]
4
[=]

21.

AL	AOS
1	1
[÷]	[÷]
6	6
[×]	[+]
3	2
[+]	[÷]
2	3
[÷]	[−]
3	3
[×]	[÷]
7	7
[−]	[=]
3	
[÷]	
7	
[=]	

22.

AL	AOS
1	3
[÷]	[+]
8	1
[×]	[÷]
5	8
[+]	[+]
3	4
[÷]	[+]
5	3
[+]	[÷]
3	5
[+]	[=]
4	
[=]	

23.

AL	AOS
8	8
[×]	[−]
7	2
[−]	[−]
3	3
[÷]	[÷]
7	7
[−]	[=]
2	
[=]	

24.

AL	AOS
1	1
[÷]	[×]
2	7
[+]	[+]
2	2
[×]	[×]
2	2
[×]	[×]
7	7
[÷]	[−]
6	3
[−]	[×]
3	6
[×]	[=]
6	
[=]	

Answer = 17/42

Section C

2. AL $\frac{10 - 8}{4}$ AOS $10 - \frac{8}{4}$

3. AL $18 \div 3 \times 2$ AOS $18 \div 3 \times 2$

4. AL $(5 \times 3 - 4)2$ AOS $5 \times 3 - 4 \times 2$

5. AL $\frac{7 \times 3 + 5}{2}$ AOS $7 \times 3 + \frac{5}{2}$

6. AL $\frac{6 + 12}{3} \times 2$ AOS $6 + \frac{12}{3} \times 2$

7. AL $9 \times 2 \div 3 \times 2$ AOS $9 \times 2 \div 3 \times 2$

8. AL $9 \times 2 \div 3 \div 2$ AOS $9 \times 2 \div 3 \div 2$

9. AL $(4 + 3)2 \times 5$ AOS $(4 + 3 \times 2)5$

10. AL $\frac{(5 - 3)2}{2}$ AOS $\frac{5 - 3 \times 2}{2}$

11. AL $\frac{3 \div 4 \times 2 + 1}{2} = \frac{3}{4} + \frac{1}{2}$ AOS $3 \div 4 \times 2 + 1 \div 2$

12. AL $\left[\left(\frac{3 \times 2}{4} + 5\right)\frac{4}{2} - 6\right]2 = 3 \times 2 + 5 \times 4 - 6 \times 2$

AOS $\frac{3 \times 2}{4} + \frac{5 \times 4}{2} - 6 \times 2$

Chapter 2

Section A

1. AL 8
AOS 8

2. AL 19
AOS 19

3. AL 49
AOS 49

4. AL 25
AOS 25

5. AL 11
AOS 11

6. AL 5
AOS 5

7. AL $3.6055513 = \sqrt{13}$
AOS $3.6055513 = \sqrt{13}$

8. AL 12
AOS 12

9. AL 6
AOS 6

10. AL 7
AOS 7

11. AL 8
AOS 8

12. AL 3
AOS 3

13. AL 15
AOS 15

14. AL -12
AOS -3

15. AL $5.3851648 = \sqrt{29}$
AOS $5.3851648 = \sqrt{29}$

16. AL 9
AOS 9

17. AL 5
AOS 5

18. AL 9
AOS 15

19. AL is not possible, since $\sqrt{-126}$ is not defined in calculator.
AOS 0.5

20. AL 0.5
AOS 5.75

Section B

21. $x = \dfrac{-11 \pm \sqrt{11^2 - 4(2)(-21)}}{2(2)}$

AL		*AOS*	
Largest	Smallest	Largest	Smallest
4	4	11	11
[×]	[×]	[x^2]	[x^2]
2	2	[+]	[+]
[×]	[×]	4	4
21	2	[×]	[×]
[+]	[+]	2	2
11	11	[×]	[×]
[x^2]	[x^2]	21	2
[=]	[=]	[=]	[=]
[√‾]	[√‾]	[√‾]	[√‾]
[+]	[+]	[−]	[+]
11	11	11	11
[+/−]	[=]	[=]	[=]
[÷]	[+/−]	[÷]	[+/−]
2	[÷]	2	[÷]
[÷]	2	[÷]	2
2	[÷]	2	[÷]
[=]	2	[=]	2
	[=]		[=]

22. $6x^2 + 17x + 5 = 0 \qquad x = \dfrac{-17 \pm \sqrt{17^2 - 4(6)(5)}}{2(6)}$

AL		*AOS*	
Largest	Smallest	Largest	Smallest
4	4	17	17

Continued on next page

(continued)

$\boxed{+/-}$	$\boxed{+/-}$	$\boxed{x^2}$	$\boxed{x^2}$
$\boxed{\times}$	$\boxed{\times}$	$\boxed{-}$	$\boxed{-}$
6	6	4	4
$\boxed{\times}$	$\boxed{\times}$	$\boxed{\times}$	$\boxed{\times}$
5	5	6	6
$\boxed{+}$	$\boxed{+}$	$\boxed{\times}$	$\boxed{\times}$
17	17	5	5
$\boxed{x^2}$	$\boxed{x^2}$	$\boxed{=}$	$\boxed{=}$
$\boxed{=}$	$\boxed{=}$	$\boxed{\sqrt{\ }}$	$\boxed{\sqrt{\ }}$
$\boxed{\sqrt{\ }}$	$\boxed{\sqrt{\ }}$	$\boxed{-}$	$\boxed{+}$
$\boxed{-}$	$\boxed{+}$	17	17
17	17	$\boxed{=}$	$\boxed{=}$
$\boxed{\div}$	$\boxed{=}$	$\boxed{\div}$	$\boxed{+/-}$
2	$\boxed{+/-}$	2	$\boxed{\div}$
$\boxed{\div}$	$\boxed{\div}$	$\boxed{\div}$	2
6	2	6	$\boxed{\div}$
$\boxed{=}$	$\boxed{\div}$	$\boxed{=}$	6
	6		$\boxed{=}$
	$\boxed{=}$		

23. $4x^2 + 12x + 9 = 0 \qquad x = \dfrac{-12 \pm \sqrt{12^2 - 4(4)(9)}}{2(4)}$

AL		*AOS*	
Largest	Smallest	Largest	Smallest
4	4	12	12
$\boxed{+/-}$	$\boxed{+/-}$	$\boxed{x^2}$	$\boxed{x^2}$
$\boxed{\times}$	$\boxed{\times}$	$\boxed{-}$	$\boxed{-}$
4	4	4	4
$\boxed{\times}$	$\boxed{\times}$	$\boxed{\times}$	$\boxed{\times}$
9	9	4	4

[+]	[+]	[×]	[×]
12	12	9	9
[x^2]	[x^2]	[=]	[=]
[=]	[=]	[√‾]	[√‾]
[√‾]	[√‾]	[−]	[+]
[−]	[+]	12	12
12	12	[÷]	[=]
[÷]	[=]	2	[+/−]
2	[+/−]	[÷]	[÷]
[÷]	[÷]	4	2
4	2	[=]	[÷]
[=]	[÷]		4
	4		[=]
	[=]		

24. $2x^2 + 5x - 4 = 0 \qquad x = \dfrac{-5 \pm \sqrt{5^2 - 4(2)(-4)}}{2(2)}$

AL		*AOS*	
Largest	Smallest	Largest	Smallest
4	4	5	5
[×]	[×]	[x^2]	[x^2]
2	2	[+]	[+]
[×]	[×]	4	4
4	4	[×]	[×]
[+]	[+]	2	2
5	5	[×]	[×]
[x^2]	[x^2]	4	4
[=]	[=]	[=]	[=]
[√‾]	[√‾]	[√‾]	[√‾]
[−]	[+]	[−]	[+]

Continued on next page

(continued)

5	[5]	5	5
[÷]	[=]	[=]	[=]
2	[+/−]	[÷]	[+/−]
[÷]	[÷]	2	[÷]
2	[2]	[÷]	2
[=]	[÷]	2	[÷]
	2	[=]	2
	[=]		[=]

25. $x^2 - 3x + 1 = 0$ $\qquad x = \dfrac{3 \pm \sqrt{3^2 - 4(1)}}{2}$

AL		*AOS*	
Largest	Smallest	Largest	Smallest
3	3	3	3
[x^2]	[x^2]	[x^2]	[x^2]
[−]	[−]	[−]	[−]
4	4	4	4
[=]	[=]	[=]	[=]
[√‾]	[√‾]	[√‾]	[√‾]
[+]	[+/−]	[+]	[+/−]
3	[+]	3	[+]
[÷]	3	[=]	3
2	[÷]	[÷]	[=]
[=]	2	2	[÷]
	[=]	[=]	2
			[=]

26. $2x^2 + 5x + 4 = 0$ $\qquad x = \dfrac{-5 \pm \sqrt{3^2 - 4(2)(4)}}{2(2)}$

AL		*AOS*	
Largest	Smallest	Largest	Smallest
4	4	5	5
[+/−]	[+/−]	[x^2]	[x^2]
[×]	[×]	[−]	[−]
2	2	4	4
[×]	[×]	[×]	[×]
4	4	2	2
[+]	[+]	[×]	[×]
5	5	4	4
[x^2]	[x^2]	[=]	[=]
[=]	[=]	[$\sqrt{\ }$]	[$\sqrt{\ }$]
[$\sqrt{\ }$]	[$\sqrt{\ }$]		
STOP—CALCULATOR SHOWS ERROR, $\sqrt{-7}$		STOP—CALCULATOR SHOWS ERROR, $\sqrt{-7}$	

27. $x^2 - x + 1 = 0 \qquad x = \dfrac{1 \pm \sqrt{1^2 - 4(1)}}{2(1)}$

AL		*AOS*	
Largest	Smallest	Largest	Smallest
1	1	1	1
[x^2]	[x^2]	[x^2]	[x^2]
[−]	[−]	[−]	[−]
4	4	4	4
[=]	[=]	[=]	[=]
[$\sqrt{\ }$]	[$\sqrt{\ }$]	[$\sqrt{\ }$]	[$\sqrt{\ }$]
STOP—CALCULATOR SHOWS ERROR, $\sqrt{-3}$		STOP—CALCULATOR SHOWS ERROR, $\sqrt{-3}$	

28. $5.39x^2 + 3.1x - 0.27 = 0$

$$x = \frac{-3.1 \pm \sqrt{(3.1)^2 - 4(5.39)(-0.27)}}{2(5.39)}$$

AL		*AOS*	
Largest	Smallest	Largest	Smallest
4	4	3.1	3.1
[×]	[×]	[x^2]	[x^2]
5.39	5.39	[+]	[+]
[×]	[×]	4	4
0.27	0.27	[×]	[×]
[+]	[+]	5.39	5.39
3.1	3.1	[×]	[×]
[x^2]	[x^2]	0.27	0.27
[=]	[=]	[=]	[=]
[√‾]	[√‾]	[√‾]	[√‾]
[−]	[+]	[−]	[+]
3.1	3.1	3.1	3.1
[÷]	[=]	[=]	[=]
2	[+/−]	[÷]	[+/−]
[÷]	[÷]	2	[÷]
5.39	2	[÷]	2
[=]	[÷]	5.39	[÷]
	5.39	[=]	5.39
	[=]		[=]

29. (21) Largest: $x = 1.5$ Smallest: $x = -7$

(22) Largest: $x = -0.333$ $(= -1/3)$ Smallest: $x = -2.5$

(23) $x = -1.5$ Largest = Smallest

(24) Largest: $x = 0.6374586$ Smallest: $x = -3.1374586$

(25) Largest: $x = 2.618034$ Smallest: $x = 0.381966$

(26) No real values

(27) No real values

(28) Largest: $x = 0.0768327$ Smallest: $x = -0.6519719$

(26) and (27):

(26) $2x^2 + 5x + 4 = 0$

$$x = \frac{-5 \pm \sqrt{5^2 - 4(2)(4)}}{2(2)} = \frac{-5 \pm i\sqrt{-[5^2 - 4(2)(4)]}}{2(2)}$$

$$= \frac{-5}{2(2)} \pm \frac{i\sqrt{-[5^2 - 4(2)(4)]}}{2(2)} = r \pm si$$

AL		AOS	
4	5	5	5
[×]	[+/−]	[x^2]	[+/−]
2	[÷]	[−]	[÷]
[×]	2	4	2
4	[÷]	[×]	[÷]
[+/−]	2	2	2
[+]	[=]	[×]	[=]
5	Display is r	4	Display is r
[x^2]		[=]	
[=]		[+/−]	
[+/−]		[√]	
[√]		[÷]	
[÷]		2	
2		[÷]	
[÷]		2	
2		[=]	
[=]		Display is s	
Display is s			

(27) $x^2 - x + 1 = 0$

$$x = \frac{-(-1) \pm \sqrt{(+1)^2 - 4(1)(1)}}{2(1)}$$

$$= \frac{-(-1) \pm i\sqrt{-[1^2 - 4(1)(1)]}}{2(1)}$$

$$= \frac{-(-1)}{2(1)} \pm \frac{i\sqrt{-[1^2 - 4(1)(1)]}}{2(1)} = r \pm si$$

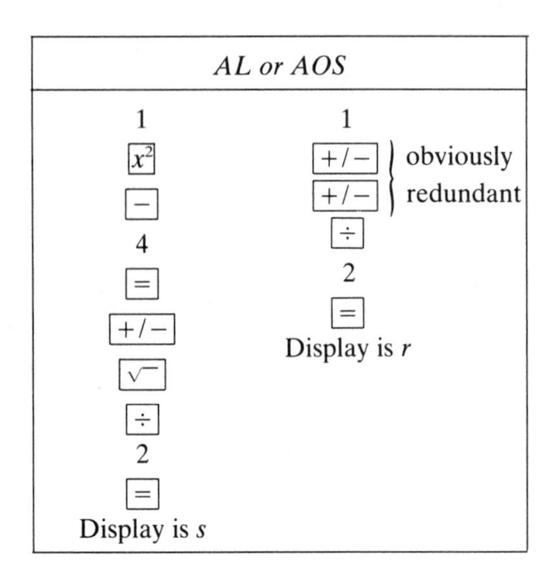

Section C

30.

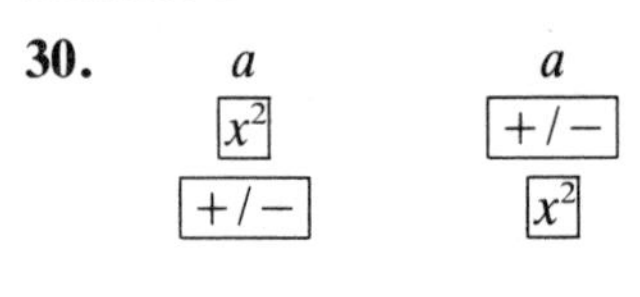

Unless $a = 0$, both sides are always different. Number on left is always negative, since x^2 gives a positive value for any a. Number on right is always positive, because the x^2 function is the last keystroke.

31. a [x^2] [$1/x$] a [$1/x$] [x^2]

Always equal since $1/x^2 = (1/x)^2$, but undefined (error) if a starts out as zero.

32. a [$1/x$] [+/−] a [+/−] [$1/x$]

For $a = 0$, undefined (error).
For all other values, they are equal since $-(1/x) = 1/(-x)$.

33. a [+/−] [√‾] a [√‾] [+/−]

If $a = 0$, equal.
For all other values *one* of the sides is undefined.
If $a > 0$, left side $= \sqrt{}$ of a negative number.
If $a < 0$, right side $= \sqrt{}$ of a negative number.

34. a [√‾] [$1/x$] a [$1/x$] [√‾]

a must be > 0, otherwise error.
If $a = 0$, the reciprocal of 0 is undefined.
If $a < 0$, $\sqrt{}$ of a negative number is imaginary.

For $a > 0$, the two expressions are equal, $1/\sqrt{x} = \sqrt{1/x}$.

35.

a	a
$\boxed{\sqrt{\ }}$	$\boxed{x^2}$
$\boxed{x^2}$	$\boxed{\sqrt{\ }}$

If $a \geq 0$, both sides are equal (to a).
If $a < 0$, left side gives an error indication but right side gives $|a|$.

Section D

36.

a	a
$\boxed{x^2}$	$\boxed{+}$
$\boxed{+}$	b
b	$\boxed{=}$
$\boxed{x^2}$	$\boxed{x^2}$
$\boxed{=}$	

Different unless either a or b is 0.
Left side is $a^2 + b^2$.
Right side is $(a + b)^2 = a^2 + 2ab + b^2$.

37.

a	a
$\boxed{x^2}$	$\boxed{-}$
$\boxed{-}$	b
b	$\boxed{=}$
$\boxed{x^2}$	$\boxed{x^2}$
$\boxed{=}$	

Different unless $b = 0$ or $b = a$.
Even with $a = 0$ and $b \neq 0$, the two sides have different values.
Left side: $a^2 - b^2$.
Right side: $(a - b)^2 = a^2 - 2ab + b^2$.

$$\begin{aligned} a^2 - b^2 &= a^2 - 2ab + b^2 \\ 0 &= -2ab + 2b^2 \\ &= 2b(-a + b) \end{aligned}$$

38.

a	a
$\boxed{x^2}$	$\boxed{\times}$
$\boxed{\times}$	b
b	$\boxed{=}$
$\boxed{x^2}$	$\boxed{x^2}$
$\boxed{=}$	

Left side: a^2b^2
Right side: $(ab)^2$
Always equal.

39.

a	a
$\boxed{+/-}$	$\boxed{+}$
$\boxed{+}$	b
b	$\boxed{=}$
$\boxed{+/-}$	$\boxed{+/-}$
$\boxed{=}$	

Left side: $-a + (-b) = -a - b$
Right side: $-(a + b) = -a - b$
Always equal for any a and b

40.

a	a
[+/−]	[−]
[−]	b
b	[=]
[+/−]	[+/−]
[=]	

Left side: $-a - (-b) = b - a$
Right side: $-(a - b) = -a + b$
Always equal for any a and b.

41.

a	a
[+/−]	[×]
[×]	b
b	[=]
[+/−]	[+/−]
[=]	

Left side: $(-a)(-b) = ab$
Right side: $-(ab) = -ab$
$ab = -ab$ only if $ab = 0$ (i.e., $a = 0$ or $b = 0$)
For all (other) values the left is the negative of the right.

42.

a	a
[1/x]	[+]
[+]	b
b	[=]
[1/x]	[1/x]
[=]	

Left side: $1/a + 1/b$

Right side: $\dfrac{1}{a + b}$

Not equal in general.

If $\dfrac{1}{a} + \dfrac{1}{b} = \dfrac{1}{a + b}$ (not defined if either a or b is zero),

then

$$b(a + b) + a(a + b) = ab$$
$$ab + b^2 + a^2 + ab = ab$$
$$a^2 + ab + b^2 = 0$$

$$a = \frac{-b \pm \sqrt{b^2 - 4(b^2)}}{2}$$

$$= \frac{-b \pm |b|\sqrt{-3}}{2}.$$ No solutions.

Thus, *never* equal.

43.

a	a
[1/x]	[×]
[×]	b
b	[=]
[1/x]	[1/x]
[=]	

Left side: $\dfrac{1}{a}\dfrac{1}{b}$

Right side: $\dfrac{1}{(ab)}$

Equal for all values of a and b but not defined if either a or b is zero.

44.

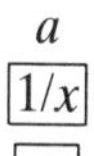

a [1/x] [÷] b [1/x] [=]

a [÷] b [=] [1/x]

Left side: $\frac{1/a}{1/b} = \frac{b}{a}$

Right side: $\frac{1}{a/b} = \frac{b}{a}$

Not defined if either a or b is zero.
Equal for all other values of a and b.

45. $\sqrt{}$ is distributive over multiplication and division.

a [$\sqrt{}$] [×] b [$\sqrt{}$] [=]

a [×] b [=] [$\sqrt{}$]

a [$\sqrt{}$] [÷] b [$\sqrt{}$] [=]

a [÷] b [=] [$\sqrt{}$]

$\sqrt{a}\sqrt{b} = \sqrt{ab}$ $\qquad \frac{\sqrt{a}}{\sqrt{b}} = \sqrt{\frac{a}{b}}$

$1/x$ is distributive over multiplication and division.

a [1/x] [×] b [1/x] [=]

a [×] b [=] [1/x]

a [1/x] [÷] b [1/x] [=]

a [÷] b [=] [1/x]

$\frac{1}{a}\frac{1}{b} = \frac{1}{ab}$ $\qquad \frac{1/a}{1/b} = \frac{b}{a} = \frac{1}{a/b} = \frac{b}{a}$

+/− is distributive over addition and subtraction.

a [+/−] [+] b [+/−] [=]

a [+] b [=] [+/−]

a [+/−] [−] b [+/−] [=]

a [−] b [=] [+/−]

$(-a) + (-b) = -(a + b)$

$-a - (-b) = b - a$ $\qquad -(a - b) = b - a$

x^2 is distributive over multiplication and division.

a	a	a	a
$\boxed{x^2}$	$\boxed{\times}$	$\boxed{x^2}$	$\boxed{\div}$
$\boxed{\times}$	b	$\boxed{\div}$	b
b	$\boxed{=}$	b	$\boxed{=}$
$\boxed{x^2}$	$\boxed{x^2}$	$\boxed{x^2}$	$\boxed{x^2}$
$\boxed{=}$		$\boxed{=}$	

$(a^2)(b^2) = (ab)^2$ $\qquad$ $(a^2)/(b^2) = (a/b)^2$

None of the trigonometric functions are distributive over any of the four operations, that is, $\sin(a + b) \neq \sin a + \sin b$.

Logarithmic functions are not distributive, that is, $\log(ab) \neq (\log a)(\log b)$.

Section E

46. If $a = 0$, error. Otherwise, values alternate, e.g., 2, 0.5, 2, 0.5, 2, 0.5, etc.

47. Values alternate between positive and negative, e.g., 2, -2, 2, -2, 2, -2, etc.

48. If $|a| > 1$, the values get larger without bound.
If $|a| < 1$, the values approach zero.
If $|a| = 1$, the values remain constant at 1.

49. If $a < 0$, error signal.
If $a = 1$, values are constant at 1.
If $a > 1$, values approach 1, that is, get smaller toward 1.
If $0 < a < 1$, values get larger, approaching 1.

50. For $a = 6$, values (recorded after $\sqrt{\ }$) approach 3.
2.45, 2.91, 2.98, 2.9974, 2.99956
For $a = 12$, values (recorded after $\sqrt{\ }$) approach 4.
3.464, 3.932, 3.9915, 3.9989
Notice $f(a) \to b$, where $a + b = b^2$.

Solving for b (in terms of a):

$$b^2 - b - a = 0$$

$$b = \frac{1 \pm \sqrt{1 + 4a}}{2}$$

All values approach $\frac{1 + \sqrt{1 + 4a}}{2}$.

51. Approaches a.
For example, if $a = 7$, recording values after each $\sqrt{}$, 2.6, 4.3, 5.49, 6.20, 6.59, 6.79, 6.89, 6.95, 6.97, 6.9867, 6.9933, 6.99667, 6.99833

Section F

52. $3\sqrt{17} - 4\sqrt{5}$

AL	*AOS*
3	3
[×]	[×]
17	17
[√‾]	[√‾]
[÷]	[−]
4	4
[−]	[×]
5	5
[√‾]	[√‾]
[×]	[=]
4	
[=]	

53. $3^2 + 4^2 + 7^2$

AL or AOS
3
[x^2]
[+]
4
[x^2]
[+]
7
[x^2]
[=]

54. $15 - 3 \times 2$

AL or AOS	*AOS only*
3	*AOS* only
[+/−]	15
[×]	[−]
2	3
[+]	[×]
15	2
[=]	[=]

55. $\dfrac{2(9)}{5 + 3}$

AL or *AOS* or *AOS*		*AL*
5	5	5
[+]	[+]	[+]
3	3	3
[=]	[=]	[÷]
[1/x]	[÷]	2
[×]	2	[÷]
2	[÷]	9
[×]	9	[=]
9	[=]	[1/x]
[=]	[1/x]	

56. In degree mode: 30 [cos]

57. In degree mode: 30 [cos] [1/x]

58. In radian mode: 3 [×] [π] [÷] 4 [=] [sin]

59. In radian mode: 3 [×] [π] [÷] 4 [=] [sin] [1/x]

60. 0.3 [inv] [sin] Answer in degrees, radians, or gradians, depending on calculator setting.

61. $\csc^{-1}(3.5) = \theta$

$3.5 = \csc\theta$

$\frac{1}{3.5} = \sin\theta$

3.5 [1/x] [inv] [sin]

Chapter 3

1.

	Display	*Memory*
	3	12
[CM]	3	0
[M+]	3	15
[M−]	3	9
[STO]	3	3
[RCL]	12	12
[PROD]	3	36
[QUO]	3	4
[EXC]	12	3
[CE]	0	12
[+/−]	−3	12

2.

	Display	*Memory*
	15	3
[CM]	15	0
[M+]	15	18
[M−]	15	−12
[STO]	15	15
[RCL]	3	3
[PROD]	15	45
[QUO]	15	0.2
[EXC]	3	15
[CE]	0	3
[+/−]	−15	3

3.

12 [STO] 3 or 12 [STO] [−] 9 [=] or 12 [STO] [÷] 4 [=] or many other possibilities

4.

3 [+/−] [STO] 5 or [CM] 3 [M−] 5 or 2 [STO] 5 [M−] or many other possibilities

5.

Keys		
a	7.3125	10.549383
[STO]		3.2479814
[√]	2.7041635	
[+]		10.560481
[RCL]		3.2496894
[=]	10.016663	10.562189
[√]	3.1649113	3.2499522
[+]		
[RCL]		10.562452
[=]	10.477411	3.2499926
[√]	3.236883	
etc.		10.562493
		3.2499989

10.562499	
3.2499998	
10.5625	Sequence converges
3.25	to 3.25

Chapter 4

Section A

1.	AL	512	**2.**	AL	512	**3.**	AL	9
	AOS	−12		AOS	1		AOS	9
4.	AL	36	**5.**	AL	36	**6.**	AL	12
	AOS	−30		AOS	12		AOS	12
7.	AL	216	**8.**	AL	61			
	AOS	6.096		AOS	−11			

9.

AL or *AOS*
5
[y^x]
3
[1/x]
[=]

10.

AL or *AOS*
5
[y^x]
3
[+/−]
[=]

11.

AL or *AOS*		
5		5
[y^x]		[y^x]
3		3
[1/x]	or	[+/−]
[+/−]		[1/x]
[=]		[=]

Section B

12.

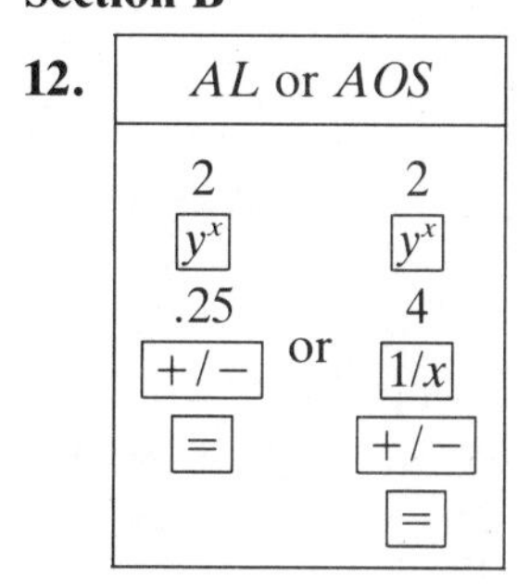

13.

AL	*AOS*
3	4
[y^x]	[×]
3	2
[=]	[−]
[STO]	3
4	[y^x]

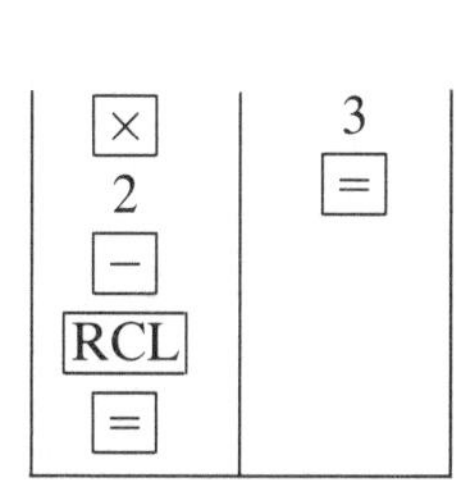

14.

AL	*AOS*
4	4
[×]	[×]
2	2
[−]	[−]
3	3
[y^x]	[=]
3	[y^x]
[=]	3
	[=]

15.

AL	*AOS*
3	4
[y^x]	[×]
3	2
[=]	[y^x]
[STO]	3
2	[−]
[y^x]	3
3	[y^x]
[×]	3
4	[=]
[−]	
[RCL]	
[=]	

16.

AL	*AOS*
3	2
[y^x]	[×]
2	5
[×]	[−]
4	4
[=]	[×]
[STO]	3
2	[y^x]
[×]	2
5	[=]
[−]	
[RCL]	
[=]	

17.

AL or *AOS*
4
[+]
7
[=]
[x^2]
[+/−]
[+]
3
[x^2]
[×]
5
[x^2]
[=]

18.

AL	*AOS*
3	4
[×]	[×]
7	5
[=]	[−]
[STO]	3
4	[×]
[×]	7
5	[=]
[−]	[STO]
[RCL]	4
[=]	[×]
[STO]	7
4	[−]
	3
[×]	[×]
7	5
[÷]	[=]
5	[÷]
[−]	[RCL]
3	[=]
[×]	
5	
[÷]	
[RCL]	
[=]	

Note use of algorithm developed on page 2 of text. Since memory is already used to store denominator, the memory cannot also be used to store intermediate results in the numerator.

19.

AL	AOS	
3	3	
×	×	
5	5	
=	−	
STO	3	
3	×	
×	7	
7	+	
=	4	
+/−	×	
M+	7	
4	=	
×	1/x or	STO
7	×	11
=	11	×
M+	×	14
11	14	÷
×	=	RCL
14		=
÷		
RCL		
=		

20.

AL	AOS
4	4
×	×
5	5
=	−
STO	3
3	×
×	7
7	=
=	1/x
+/−	×
M+	3
3	×
×	5
5	=
÷	+/−
RCL	+
=	4
STO	×
4	7
×	=
7	
−	
RCL	
=	

21.

AL	AOS
2	6
y^x	−
3	5
×	×
5	2
+/−	y^x
+	3
6	=
×	×
3	3
+	+
5	5
=	=

22.

AL	*AOS*
3	2
y^x	×
2	7
=	−
STO	3
2	x^2
×	=
7	y^x
−	3
RCL	=
y^x	STO
3	3
=	×
STO	5
3	−
×	2
5	×
÷	4
4	=
−	x^2
2	×
×	RCL
4	=
=	STO
x^2	3
×	×
RCL	7
=	$\sqrt{\ }$
STO	+
3	5
×	=
7	y^x
$\sqrt{\ }$	2
+	+/−
5	×
y^x	RCL
2	=
+/−	
×	
RCL	
=	

23.

AL	*AOS*
4	2
y^x	×
3	8
=	−
STO	14
2	+
×	3
5	=
+	y^x
7	3
y^x	=
3	STO
−	2
RCL	×
=	5
STO	+
2	7
×	=
8	y^x
−	3
14	−
+	4
3	y^x
y^x	3
3	=
=	÷
1/x	RCL
×	=
RCL	
=	

24.

AL	*AOS*
5	37
×	−
2	5
+/−	×
+	2
37	=
=	STO
STO	4
5	+
y^x	3
RCL	×
×	5
3	y^x
+	RCL
4	=
=	

25.

AL	*AOS*
2	2
+/−	+/−
STO	STO
×	×
3	3
−	−
4	4
×	=
RCL	×
−	RCL
7	−
×	7
RCL	=
−	×
2	RCL
=	−
	2
	=

Display in calculator:

-28

26.

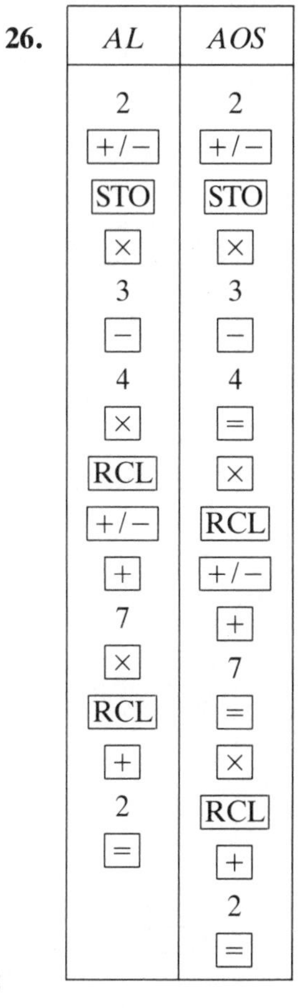

AL	*AOS*
2	2
+/−	+/−
STO	STO
×	×
3	3
−	−
4	4
×	=
RCL	×
+/−	RCL
+	+/−
7	+
×	7
RCL	=
+	×
2	RCL
=	+
	2
	=

Display in calculator:

28

27. Since direct calculation of $(-1.238)^3$ is difficult on many calculators, as was discussed on pages 31 and 32, we have chosen to rewrite the expression as

$$\frac{5x^2 - 2x + 4}{(3x^2 - 1)x + 1} \text{ or } \frac{(5x - 2)x + 4}{(3x^2 - 1)x + 1}.$$

AL	*AOS*
1.238	1.238
+/−	+/−

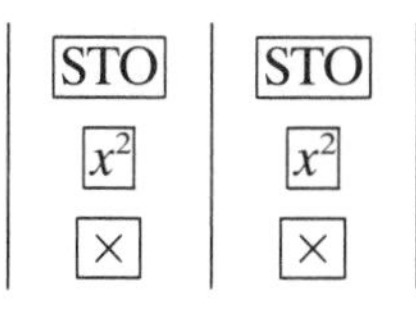

STO	STO
x^2	x^2
×	×

Continued on next page

(continued)

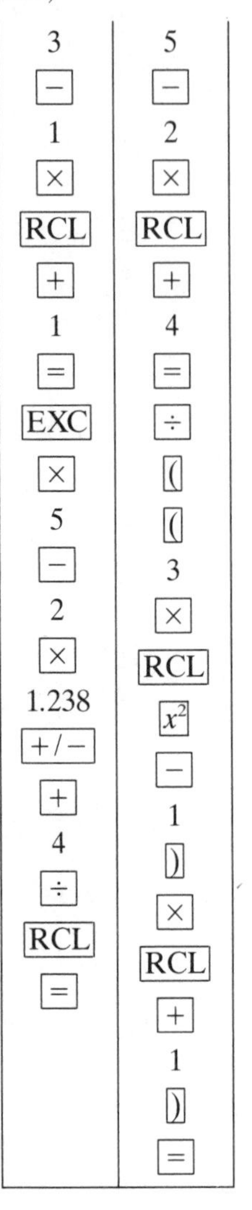

Also see discussion of this technique of nested interval decomposition in chapter 7, page 68.

Display in calculator: -4.0932943

Section C

28. $A = \$30\ 000$

$n = (20 \text{ years})(12 \text{ months/year}) = 240$

$$i = \frac{(0.14 \text{ per year})}{12 \text{ months/year}} = \frac{0.14}{12}$$

Need to calculate $R = \dfrac{(30\ 000)(0.14/12)}{1 - (1 + 0.14/12) - 240}$

AL	*AOS*
.14	.14
÷	÷
12	12
=	=
STO	STO
+	+
1	1
y^x	=
240	y^x
+/−	240
=	+/−
+/−	=
+	+/−
1	+
÷	1
30000	=
÷	÷
RCL	30000
=	÷
1/x	RCL
	=
	1/x

$R =$ \$373.05624 or \$373.06. Over the life of the loan, the borrower pays \$89 533.50 or \$89 534.40 depending whether R is rounded before or after multiplying by 240.

29. $R = \$186.53$ $\qquad i = \dfrac{0.14}{(365/14)} = \dfrac{(0.14)(14)}{365}$

$$n = \frac{-\log(1 - Ai / R)}{\log(1 + i)} = \frac{-\log(1 - (30\,000)i / 186.53)}{\log(1 + i)}$$

AL or AOS	
.14	14
×	÷
	365

Continued on next page

(continued)

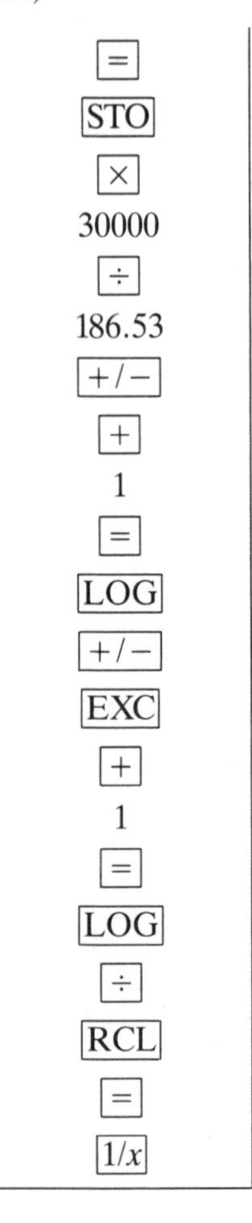

Calculation gives n = 372.04783, that is, 372 payments of $186.53 each, with the last payment (0.0478282)($186.53) = $8.92. The entire loan will be paid off in (373)(2) = 746 weeks or 14.307 years (14 years, 4 months). The total amount paid will be ($186.53)(372) + $8.92 = $69 398.08.

30. Mean = $5.66\overline{6}$ or $5\,\frac{2}{3}$

Standard deviation = 2.4267033 or 2.4

Section D

Column A 7
Column B 9
Column C 14
Column D 4

Chapter 5

Numbers in brackets, [ ], refer to exercises in chapter 4, page 48.

1. [9] $5^{1/3}$

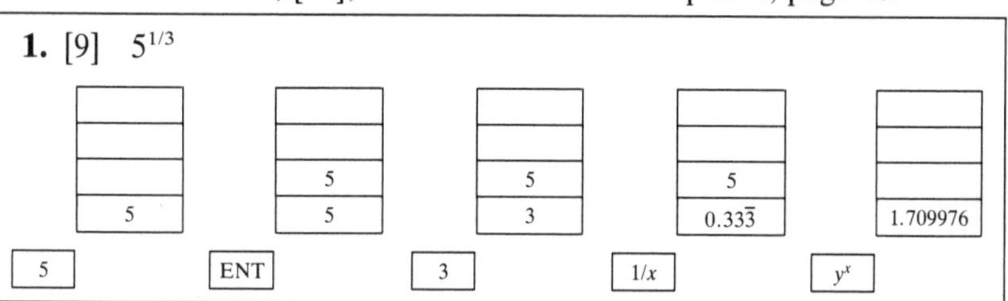

2. [10] 5^{-3}

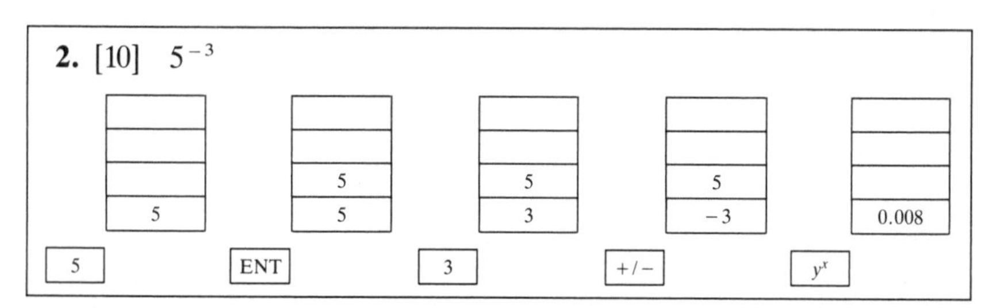

3. [11] $5^{-1/3}$

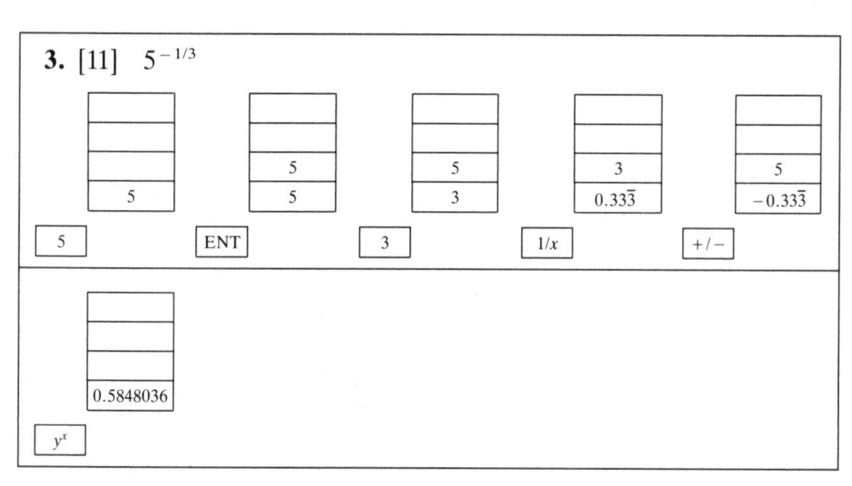

4. [12] $2^{-0.25}$

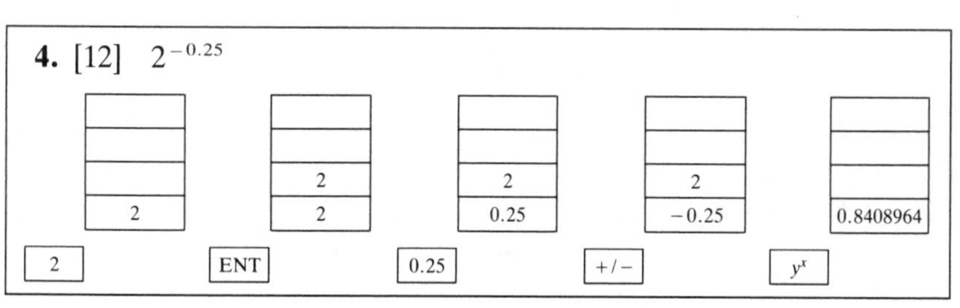

5. [13] $4(2) - 3^3$

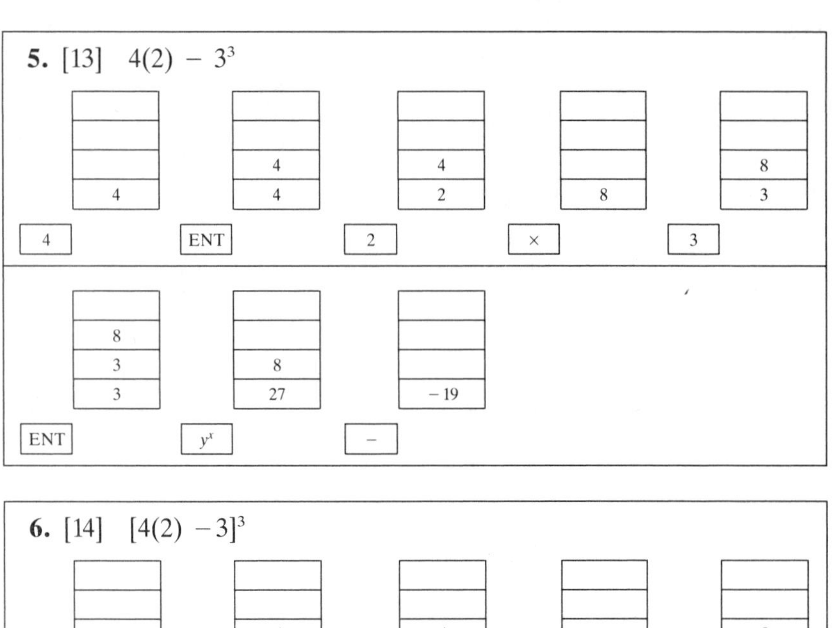

6. [14] $[4(2) - 3]^3$

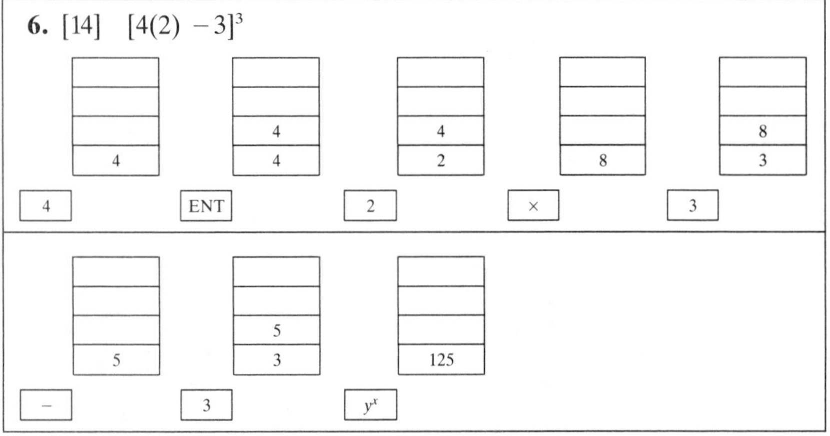

7. [15] $4(2^3) - 3^3$

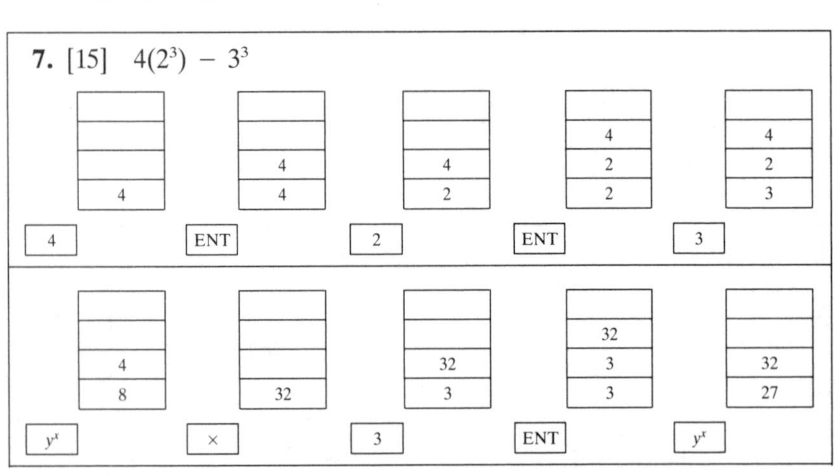

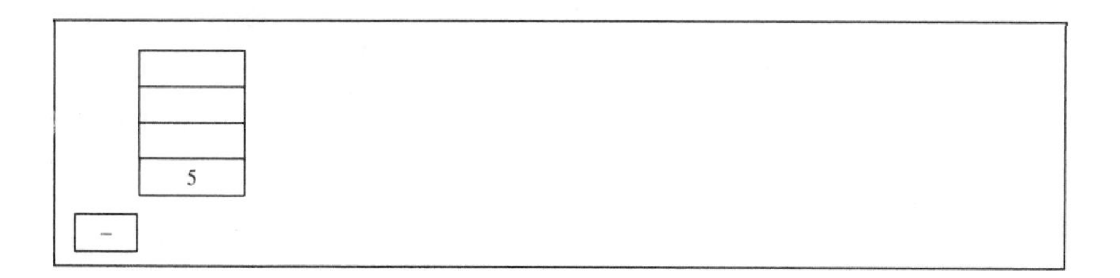

8. [16] $2 \times 5 - 4 \times 3^2$

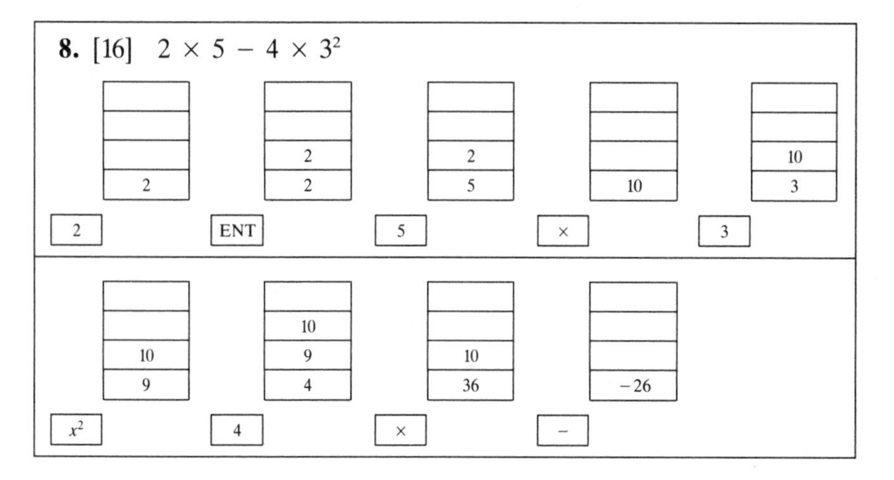

9. [17] $3^2 \times 5^2 - (4 + 7)^2$

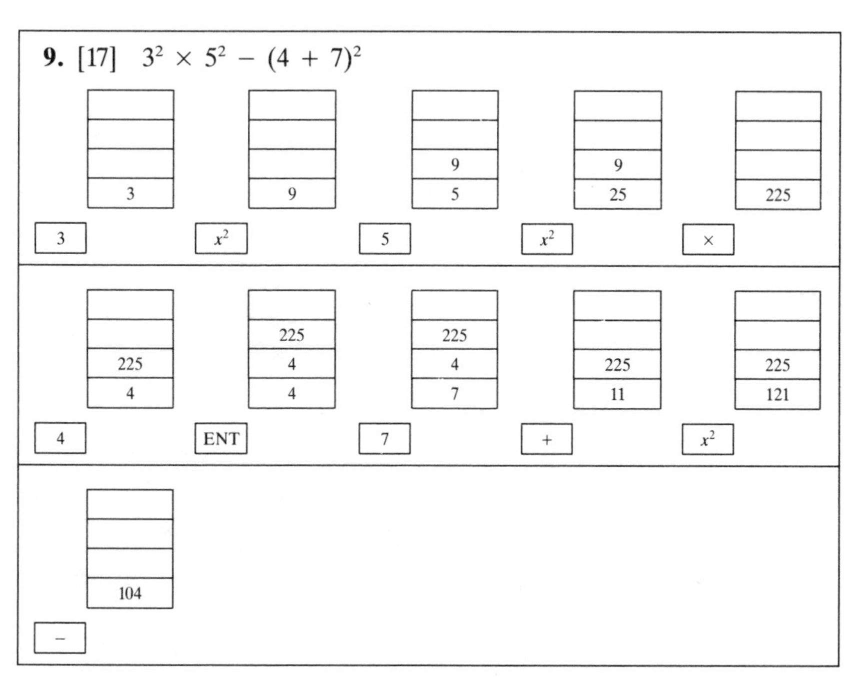

10. [18] $\dfrac{4 \times 7 - 3 \times 5}{4 \times 5 - 3 \times 7}$

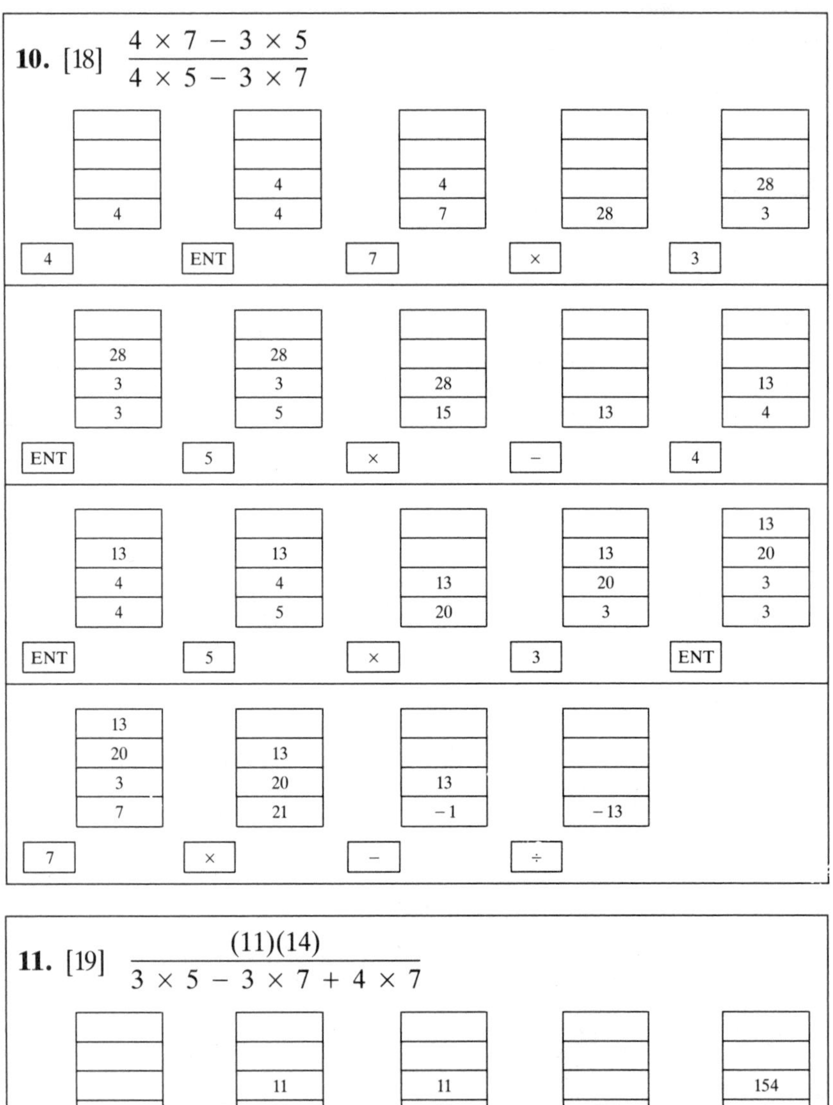

11. [19] $\dfrac{(11)(14)}{3 \times 5 - 3 \times 7 + 4 \times 7}$

Stack				
				154
154	154		154	15
3	3	154	15	3
3	5	15	3	3
ENT	5	×	3	ENT

154				154
15	154		154	−6
3	15	154	−6	4
7	21	−6	4	4
7	×	−	4	ENT

154			
−6	154		
4	−6	154	
7	28	22	7
7	×	+	÷

12. [20] $4 \times 7 - \dfrac{3 \times 5}{4 \times 5 - 3 \times 7}$

	3	3		15
3	3	5	15	4
3	ENT	5	×	4

				15
15	15		15	20
4	4	15	20	3
4	5	20	3	3
ENT	5	×	3	ENT

15				
20	15			
3	20	15		
7	21	−1	−15	15
7	×	−	÷	+/−

	15	15		
15	4	4	15	
4	4	7	28	43
4	ENT	7	×	+

13. [21] $5 + 3(6 - 5 \times 2^3)$

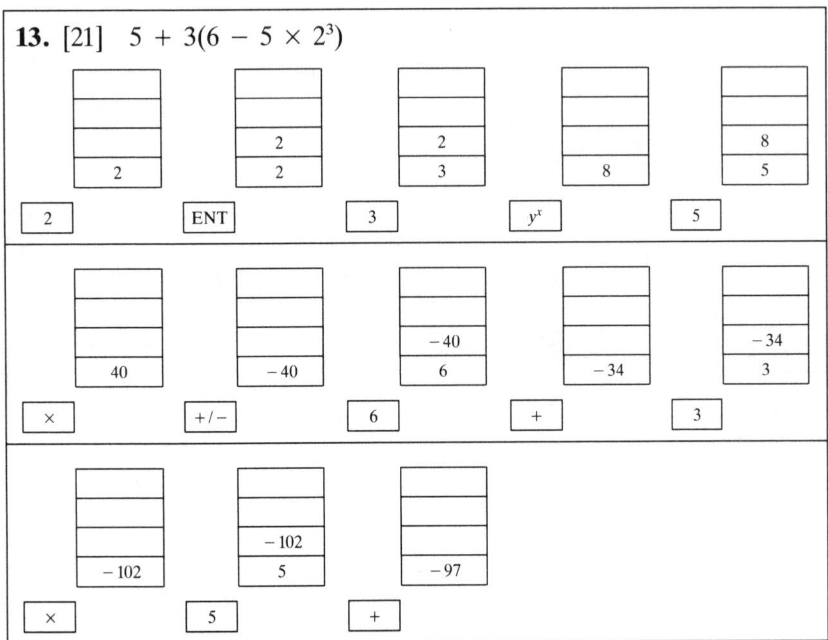

14. [22] $(2 \times 7 - 3^2)^3 (3 \times 5 - 2 \times 4)^2 (3\sqrt{7} + 5)^{-2}$

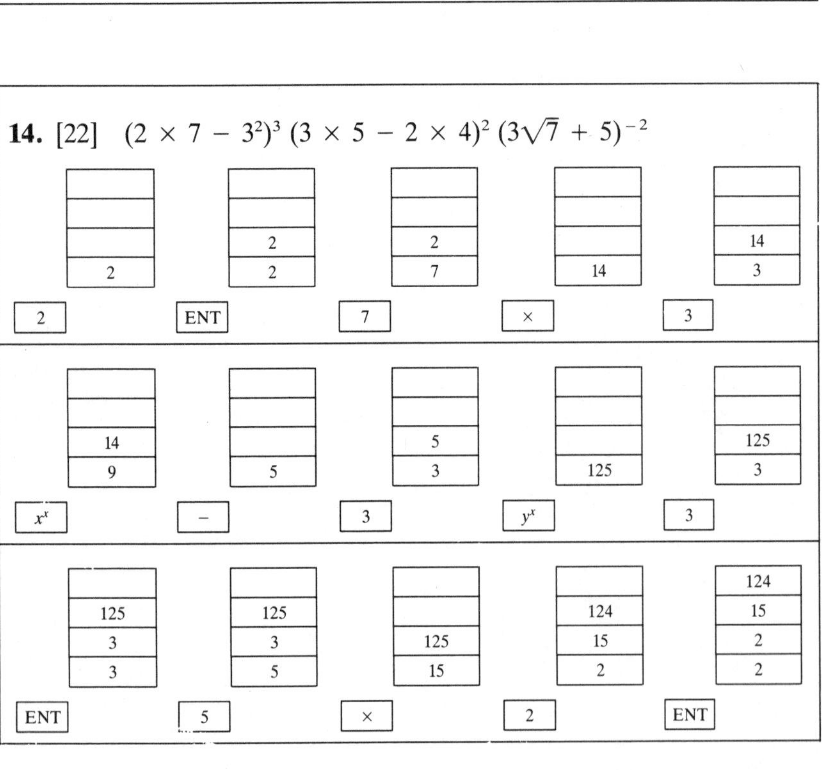

124				
15	124			
2	15	124	124	
4	8	7	49	6076
4	×	−	x^2	×

		6076		6076
6076	6076	2.6457…	6076	7.9372…
7	2.6457…	3	7.9372…	5
7	$\sqrt{}$	3	×	5

	6076	6076		
6076	29.372…	12.9372…	6076	
12.9372…	2	−2	.00597…	36.302251
+	2	+/−	y^x	×

15. [23] $\dfrac{(2 \times 5 + 7)^3 - 4^3}{(2 \times 8 - 14 + 3)^3}$

	2	2		10
2	2	5	10	7
2	ENT	5	×	7

				4913
	17		4913	4
17	3	4913	4	4
+	3	y^x	4	ENT

4913				4849
4	4913		4849	2
3	64	4849	2	2
3	y^x	−	2	ENT

Continued on next page

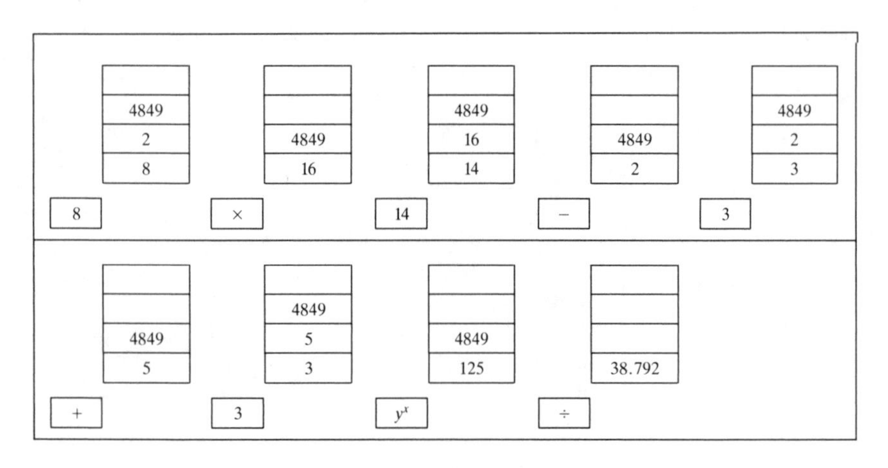

16. [24] $4 + 3 \times 5^{(37 - 5 \times 2)}$

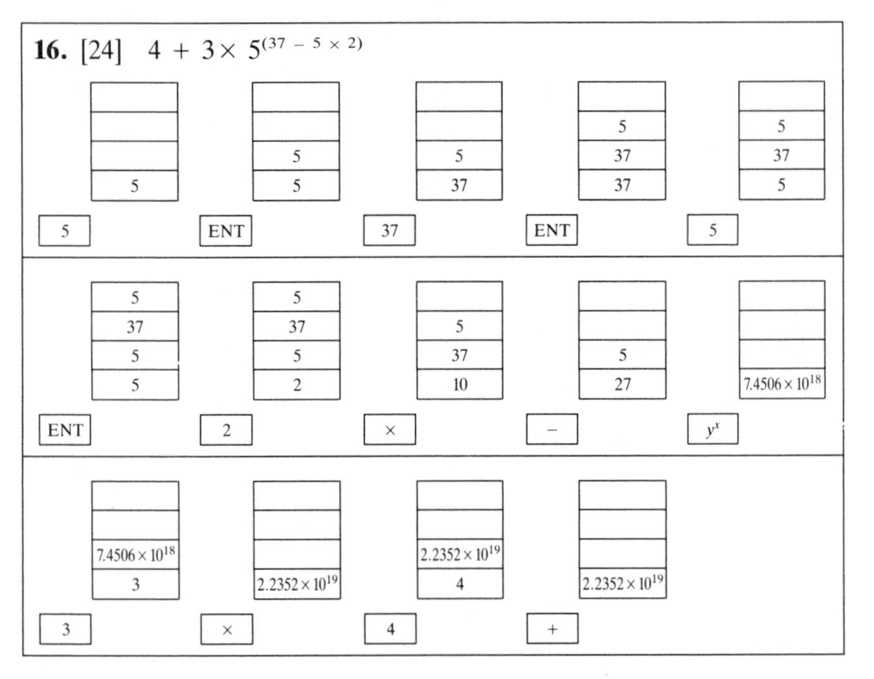

17. [25] $[(3x - 4)x - 7]x - 2$, for $x = -2$

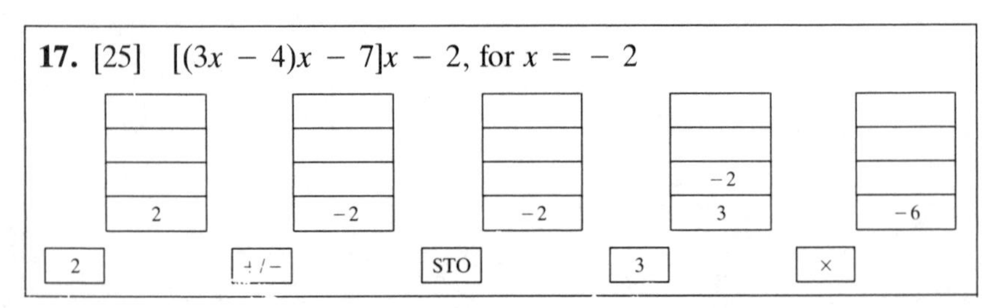

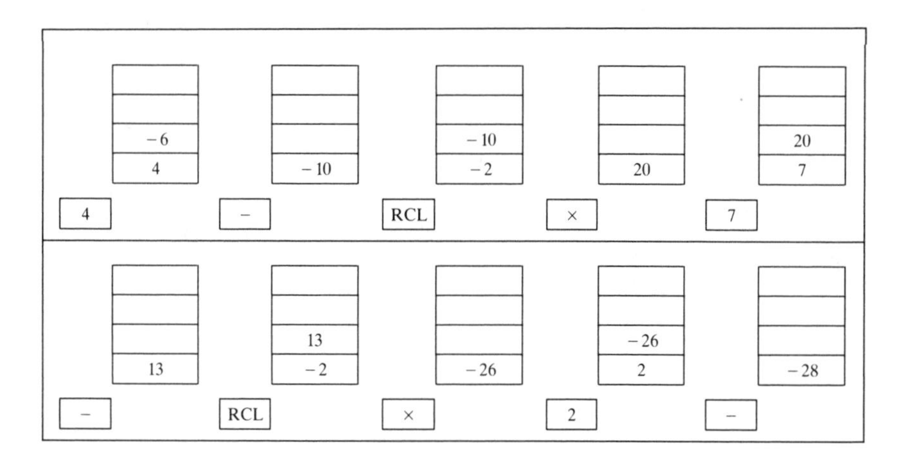

18. [26] $2 + x[7 - x(3x - 4)] = [(3x - 4)(-x) + 7]x + 2$, for $x = -2$

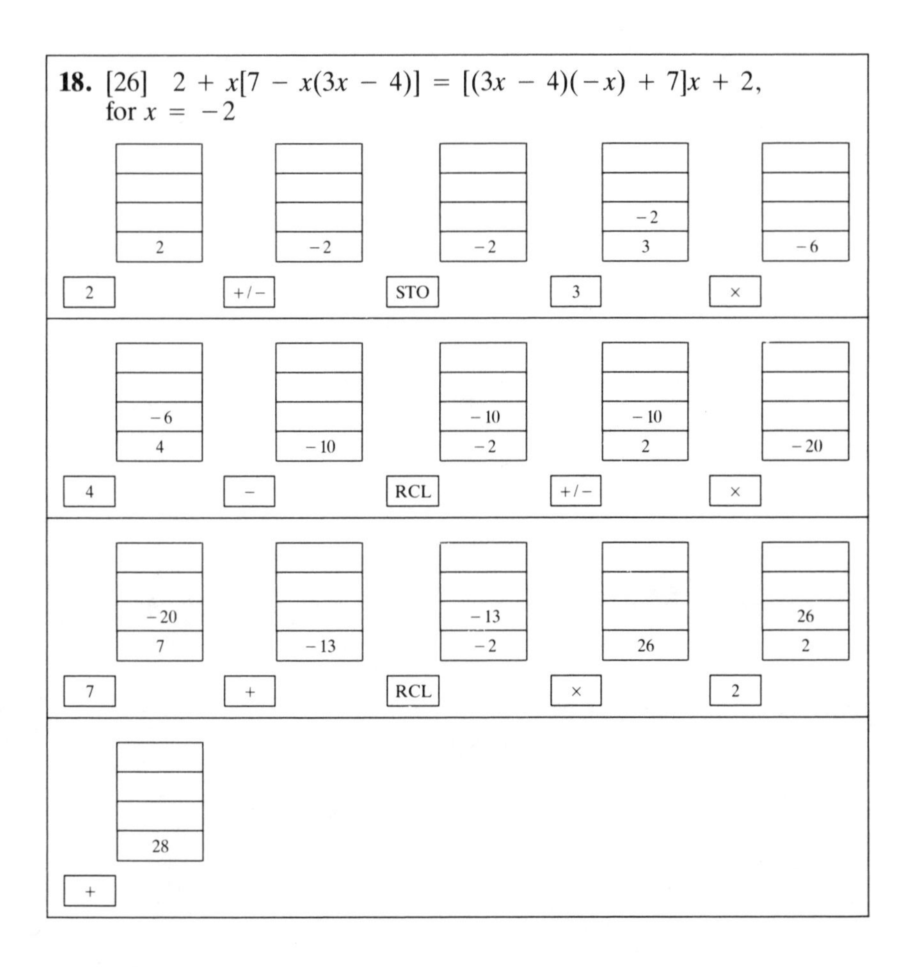

19. [27] $\dfrac{5x^2 - 2x + 4}{3x^3 - x + 1} = \dfrac{5x^2 - 2x + 4}{(3x^2 - 1)x + 1}$, for $x = -1.238$

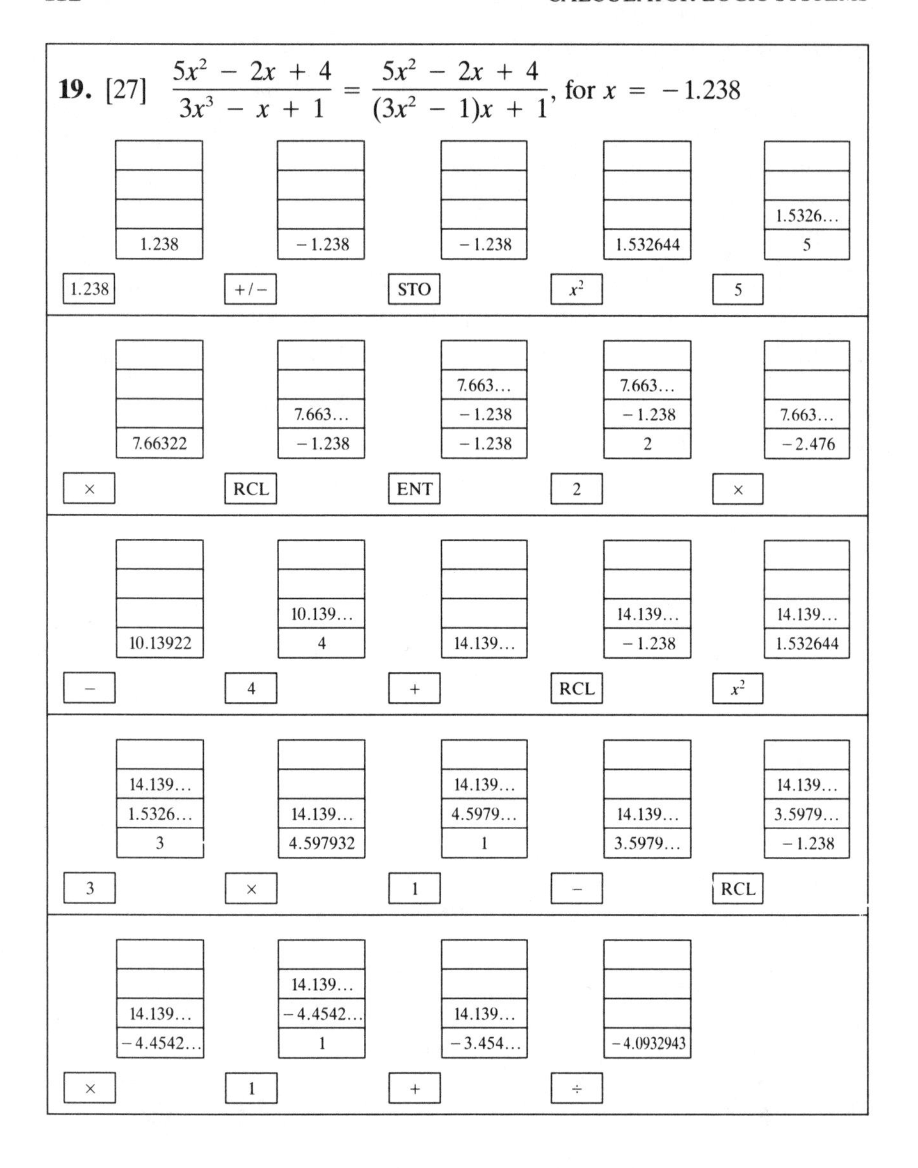

Chapter 6

Section A

1. 3.04×10^2 **2.** 8×10^7 **3.** 3×10^{-1}

4. 5.6×10^{-3} **5.** 6.49×10^0

Section B

6. 1 000 **7.** 0.000 1 **8.** 0.1
9. 1 **10.** 0.003 12 **11.** 40 500
12. 0.090 0 **13.** 230 000 **14.** 3.723

Section C

15. $0.013 \times 10^{-3} = 1.3 \times 10^{-2} \times 10^{-3} = 1.3 \times 10^{-5}$

16. $295.4 \times 10^{2} = 2.954 \times 10^{2} \times 10^{2} = 2.954 \times 10^{4}$

17. $3.75 \times 10^{3} = 3.75 \times 10^{4} \times 10^{-1} = 37\,500 \times 10^{-1}$

18. $6.03 \times 10^{-2} = 6.03 \times 10^{-1} \times 10^{-1} = 0.603 \times 10^{-1}$

19. $0.6 \times 10^{-2} = 0.6 \times 10^{2} \times 10^{-4} = 60 \times 10^{-4}$

20. $63.75 \times 10^{6} = 63.75 \times 10^{3} \times 10^{3} = 63\,750 \times 10^{3}$

Section D

21. $$\frac{(3 \times 10^{7})(3 \times 10^{-4})}{6 \times 10^{3}} = \frac{3 \times 3}{6} \times 10^{7} \times 10^{-4} \times 10^{-3} = 1.5 \times 10^{0} = 1.5$$

22. $$\frac{(16 \times 10^{-3})(3 \times 10^{4})}{1.2 \times 10^{-6}} = \frac{16}{0.4} \times 10^{-3} \times 10^{4} \times 10^{6} = 40 \times 10^{7} = 4 \times 10^{1} \times 10^{7} = 4 \times 10^{8}$$

23. $$\frac{(3.5 \times 10^{4})(12 \times 10^{-1})}{2.1 \times 10^{-8}} = \frac{(3.5)(12)}{2.1} \times 10^{4} \times 10^{-1} \times 10^{8} = 20 \times 10^{11} = 2 \times 10^{1} \times 10^{11} = 2 \times 10^{12}$$

24. $(6 \times 10^{-25})^{2} = 36 \times 10^{-50} = 3.6 \times 10^{1} \times 10^{-50} = 3.6 \times 10^{-49}$

25. $(2 \times 10^{19})^{3} = 8 \times 10^{57}$

26. $3.07 \times 10^{4} + 3.9 \times 10^{4} = (3.07 + 3.9) \times 10^{4} = 6.97 \times 10^{4}$

27. $3 \times 10^{2} + 5 \times 10^{-2} = 300 + 0.05 = 300.05$

28. $$6 \times 10^{23} - 7 \times 10^{21} = (6 - 7 \times 10^{-2}) \times 10^{23} = (6 - 0.07) \times 10^{23} = 5.93 \times 10^{23}$$

29. $$4.5 \times 10^{151} - 7.8 \times 10^{154} = (4.5 \times 10^{-3} - 7.8) \times 10^{154} = (0.0045 - 7.8) \times 10^{154} = -7.7955 \times 10^{154}$$

30. $5.4 \times 10^{-141} + 3.7 \times 10^{-144} = (5.4 + 3.7 \times 10^{-3}) \times 10^{-141}$
$= (5.4 + 0.0037) \times 10^{-141}$
$= 5.4037 \times 10^{-141}$

31. $(6 \times 10^5)^2 (3 \times 10^4)^{-3} = 6^2 \times 10^{25} \times 3^{-3} \times 10^{-12}$
$= \dfrac{36}{27} \times 10^{13} = 1.\overline{3} \times 10^{13}$

32. $(3 \times 10^{-4})^2 (5 \times 10^7)^{-3} = 3^2 \times 10^{-8} \times 5^{-3} \times 10^{-21}$
$= \dfrac{9}{125} \times 10^{-29} = 0.072 \times 10^{-29}$
$= 7.2 \times 10^{-2} \times 10^{-29}$
$= 7.2 \times 10^{-31}$

33. $\dfrac{(237\ 000)\ (0.000\ 374)\ (305)}{(0.091)\ (5\ 100)\ (100.5)}$

$= \dfrac{2.37 \times 10^5 \times 3.74 \times 10^{-4} \times 3.05 \times 10^2}{9.1 \times 10^{-2} \times 5.1 \times 10^3 \times 1.005 \times 10^2}$

$= \dfrac{(2.37)\ (3.74)\ (3.05)}{(9.1)\ (5.1)\ (1.005)} \times 10^5 \times 10^{-4} \times 10^2 \times 10^2 \times 10^{-3} \times 10^{-2}$

$= 0.579\ 618\ 4 \times 10^0 = 5.796\ 184 \times 10^{-1}$

or 5.8×10^{-1}, to correct number of significant digits

34. $\dfrac{(589\ 060\ 000)\ (639)\ (87\ 249)}{(0.001\ 58)\ (0.003\ 4)\ (0.059\ 06)}$

$= \dfrac{5.8906 \times 10^8 \times 6.39 \times 10^2 \times 8.7249 \times 10^4}{1.58 \times 10^{-3} \times 3.4 \times 10^{-3} \times 5.906 \times 10^{-2}}$

$= \dfrac{5.8906 \times 6.39 \times 8.7249}{1.58 \times 3.4 \times 5.906} \times 10^8 \times 10^2 \times 10^4 \times 10^3 \times 10^3 \times 10^2$

$= 10.351\ 217 \times 10^{22} = 1.035\ 121\ 7 \times 10^1 \times 10^{22}$

$= 1.035\ 121\ 7 \times 10^{23}$

$= 1.0 \times 10^{23}$, correct to two significant digits

Chapter 7

Section A

1. a) $f(x) = x^3 - 3x^2 - 9x + 30 = [(x - 3)x - 9]x + 30$

b) will vary; the following is a possible set of values:

x	0	1	2	3	4	5	7
$f(x)$	30	19	8	3	10	35	163

x	-1	-2	-3	-4	-5	3.2
$f(x)$	35	28	3	-46	-125	3.248

c) $f(x) = x^3 - 3x^2 - 9x + 30$

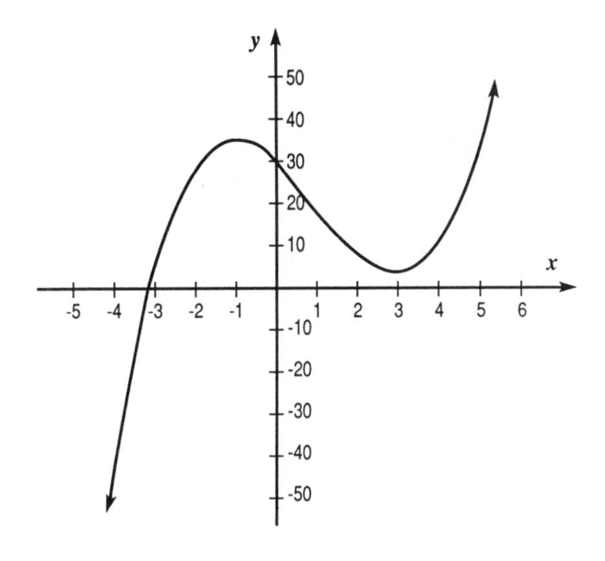

d) one solution

e) $-4 < x < -3$

f)

x	$f(x)$
-3	3
-4	-46
-3.1	-0.721
-3.08	$+0.0427$
-3.09	-0.338

$-3.09 < x < -3.08$

2. a) $f(x) = 0.3x^3 - 1.8x^2 + 1.5x + 5 = [(0.3x - 1.8)x + 1.5]x + 5$

b)

x	0	1	0.5	2	3	4	3.5
$f(x)$	5	5	5.34	3.2	1.4	1.4	1.06

x	5	6	7	−1	−2	−3	−4
$f(x)$	5	14	30.2	1.4	−7.6	−23.8	−49

c) $f(x) = 0.3x^3 - 1.8x^2 + 1.5x + 5$

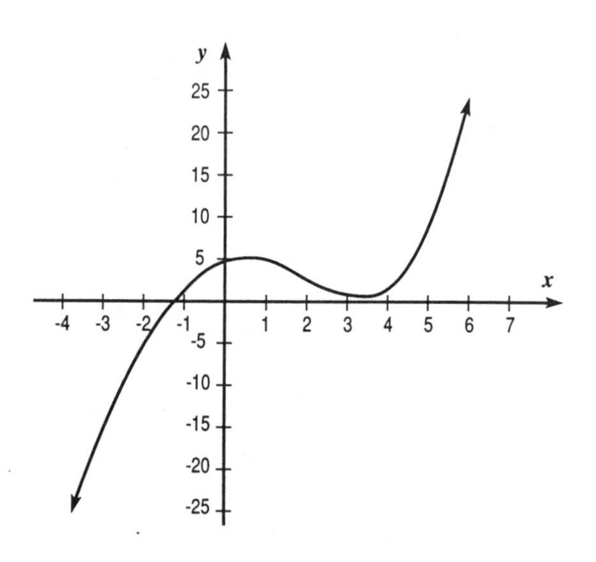

d) one solution

e) $-2 < x < -1$

f)

x	$f(x)$
−1	1.4
−2	−7.6
−1.4	−1.45
−1.2	+0.09
−1.3	−0.65
−1.23	−0.126
−1.21	+0.018
−1.22	−0.05

$-1.22 < x < -1.21$

3. a) $f(x) = 5x^3 - 26x^2 - 29x + 100 = [(5x - 26)x - 29]x + 100$

b)

x	0	1	2	3	4	5	6	7
$f(x)$	100	50	−22	−86	−112	−70	70	338

x	−1	−2	−3	−4	−2.5
$f(x)$	98	14	−182	−520	−68.125

c) $f(x) = 5x^3 - 26x^2 - 29x + 100$

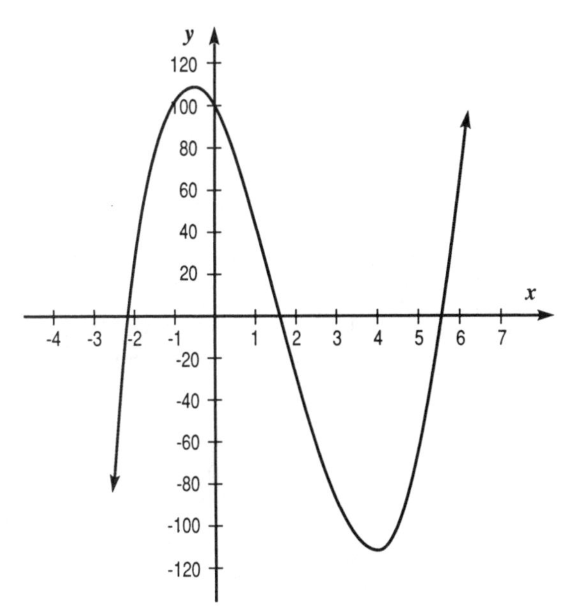

d) three solutions

e) $-3 < x < -2$ $1 < x < 2$ $5 < x < 6$

f)

x	$f(x)$
−3	−182
−2	14
−2.2	−15.38
−2.1	−0.065
−2.09	+1.393

x	$f(x)$
1	50
2	−22
1.5	14.875
1.7	+0.125
1.8	−7.28
1.71	−0.616

x	$f(x)$
5	−70
6	70
5.5	−14.12
5.7	15.925
5.6	0.32
5.59	−1.18

$-2.10 < x < -2.09$ $1.70 < x < 1.71$ $5.59 < x < 5.60$

4. a) $f(x) = x^4 - 5x^3 + 1 = (x - 5)x^2 \cdot x + 1$

b)

x	0	1	2	3	-1	-2	4	5
$f(x)$	1	-3	-23	-53	7	57	-63	1

x	4	5	6	5.3
$f(x)$	-63	1	217	45.7

Note:
If $x < 0$, then $(x-5) < 0$ and $x^3 < 0$. Therefore, $(x-5)x^3 > 0$.
If $x > 5$, then $(x-5) > 0$ and $x^3 > 0$. Therefore, $(x-5)x^3 > 0$.

c) $f(x) = x^4 - 5x^3 + 1$

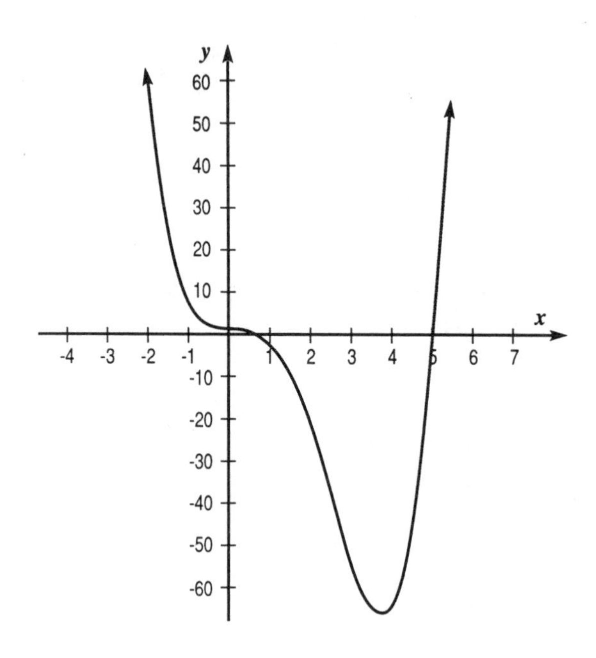

d) two solutions

e) $0 < x < 1$ $\qquad$ $4 < x < 5$

f)

x	$f(x)$
0	1
1	-3
0.5	$+0.4375$
0.7	-0.4749
0.6	$+0.0496$
0.63	-0.093
0.62	-0.04
0.61	$+0.003$

x	$f(x)$
4	-63
5	1
4.9	-10.76
4.99	-0.24

$0.61 < x < 0.62$ $\qquad$ $4.99 < x < 5.00$

5. a) $f(x) = 4x^3 - 19x^2 - 25x + 80 = [(4x - 19)x - 25]x + 80$

b)

x	0	1	2	3	4	5	6
$f(x)$	80	40	−14	−58	−68	−20	110

x	−1	−2	−3
$f(x)$	82	22	−124

c) $f(x) = 4x^3 - 19x^2 - 25x + 80$

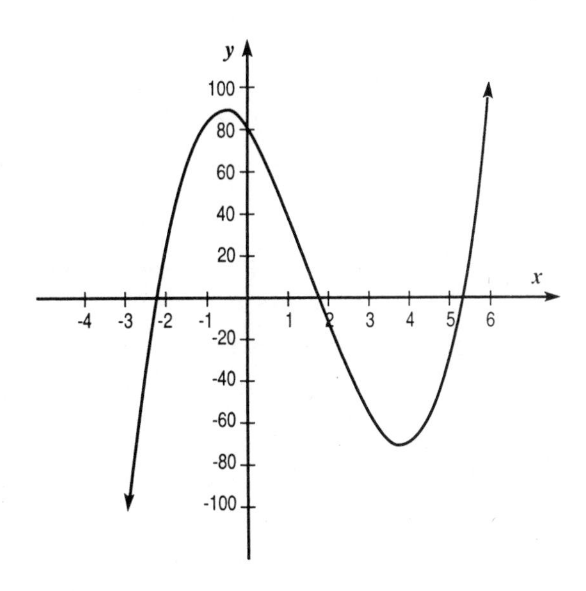

d) three solutions

e) $-3 < x < -2$ $\quad$ $1 < x < 2$ $\quad$ $5 < x < 6$

f)

x	$f(x)$
−3	−124
−2	22
−2.2	+0.448
−2.3	−11.68
−2.21	−0.723

$-2.21 < x < -2.20$

x	$f(x)$
1	40
2	−14
1.7	2.24
1.8	−3.23
1.75	−0.5
1.73	+0.596
1.74	+0.048

$1.74 < x < 1.75$

x	$f(x)$
5	−20
6	110
5.2	−1.33
5.3	9.298
5.25	3.875
5.22	+0.727
5.21	−0.305

$5.21 < x < 5.22$

6. a) $f(x) = 2x^3 + 7x^2 + 10 = (2x + 7)x^2 + 10$

b)

x	0	1	-1	-2	-3	-4	-5	-4.5
$f(x)$	10	19	15	22	19	-6	-65	-30.5

Note: If $x > 0$, then $f(x) > 0$, since all coefficients are positive.

c) $f(x) = 2x^3 + 7x^2 + 10$

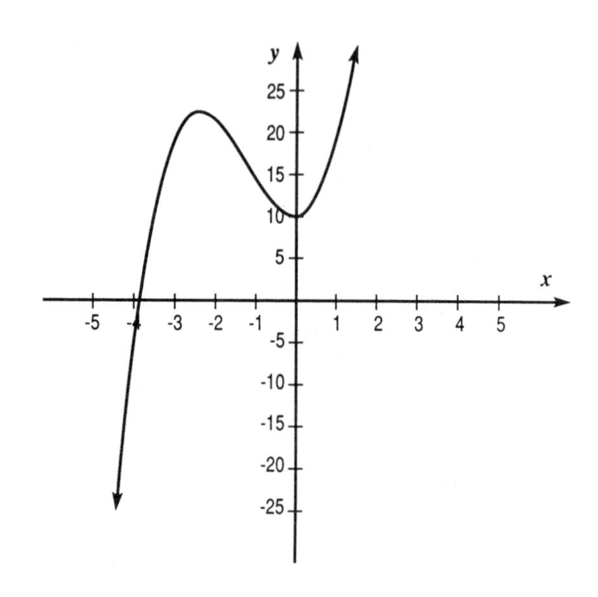

d) one solution

e) $-4 < x < -3$

f)

x	$f(x)$
-3	19
-4	-6
-3.8	1.34
-3.9	-2.17
-3.85	-0.376
-3.83	$+0.319$
-3.84	-0.027

$-3.84 < x < -3.83$

7. a) $f(x) = 4x^2 - 2x^4 - 1 = -2x^4 + 4x^2 - 1 = (-2x^2 + 4)x^2 - 1$

b)

x	0	± 1	± 2	± 3	± 0.5	± 1.5
$f(x)$	-1	1	-17	-127	-0.125	-2.125

Graph is symmetric about y-axis since $f(x) = f(-x)$. Notice that all exponents are even.

c) $f(x) = 4x^2 - 2x^4 - 1$

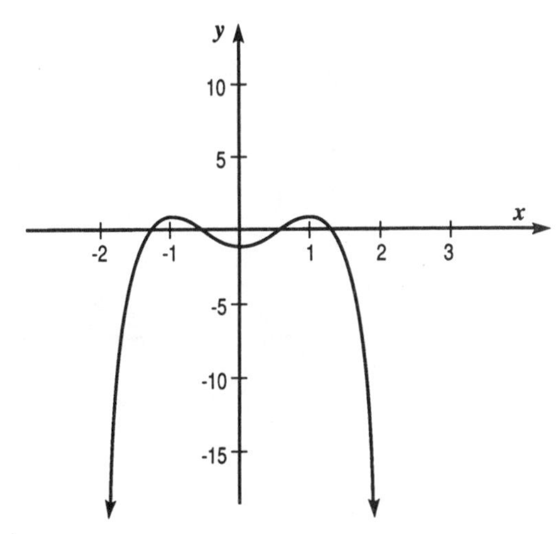

d) four solutions

e) $-2 < x < -1$ $\quad -1 < x < 0$ $\quad 0 < x < 1$ $\quad 1 < x < 2$

f)

x	$f(x)$
0	-1
1	1
0.50	-0.125
0.07	$+0.48$
0.60	$+0.181$
0.55	$+0.027$
0.53	-0.034
0.54	-0.0037

$0.54 < x < 0.55$

x	$f(x)$
1	1
2	-17
1.2	$+0.613$
1.3	$+0.478$
1.4	-0.843
1.35	-0.353
1.33	-0.182
1.32	-0.102
1.31	-0.026

$1.30 < x < 1.31$

by symmetry,

$-0.55 < x < -0.54$ $\qquad -1.31 < x < -1.30$

Section B

1. $\dfrac{1}{x^2 + 1} = 2x,\ 1 = 2x^3 + 2x,\ 2x^3 + 2x - 1 = 0$

$f(x) = 2x^3 + 2x - 1 = (2x^2 + 2)x - 1$ or $(x^2 + 1)2x - 1$

x	0	1	2	−1	−2
$f(x)$	−1	3	19	−5	−21

$f(x) = 2x^3 + 2x - 1$

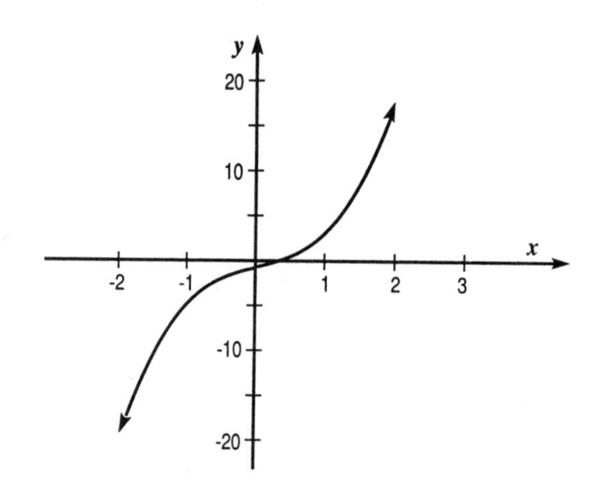

f has one solution: $0 < x < 1$

x	$f(x)$
0	−1
1	3
0.5	+0.25
0.3	−0.35
0.4	−0.072
0.42	−0.0118
0.44	+0.050
0.43	+0.019

$0.42 < x < 0.43$

2. $2 \sin x = x, x > 0$

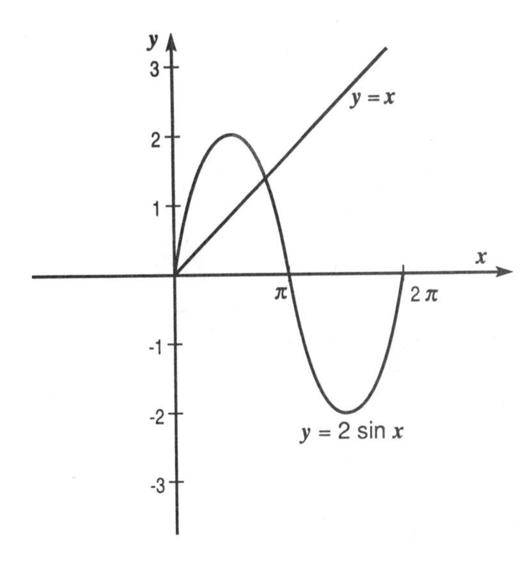

From the graphs of $y = x$ and $y = 2 \sin x$ on the same axis, the graphs appear to intersect in the neighborhood of $x = 1.8$.
Let $f(x) = 2 \sin x - x$

Note: Be sure that calculator is in radian mode!

x	$f(x)$
1.6	+0.399
2.0	−0.181
1.8	+0.148
1.9	−0.0074
1.89	+0.00897

$1.89 < x < 1.90$

3. $\sin x = x^2$

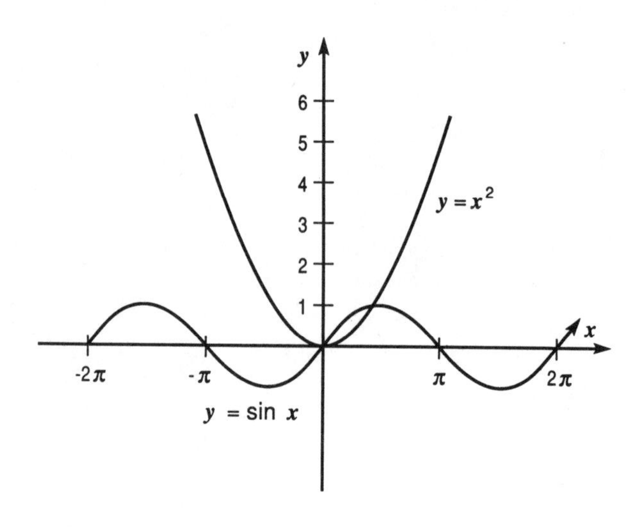

$x = 0$ is an exact solution.

From the graphs of $y = x^2$ and $y = \sin x$ on the same axis, the graphs appear to intersect in the region of $x = +0.8$.

Let $f(x) = \sin x - x^2$

Note: Be sure that calculator is in radian mode!

x	$f(x)$
0.6	+0.205
1.0	−0.159
0.8	+0.077
0.9	−0.027
0.85	+0.029
0.87	+0.0074
0.88	−0.0037

$0.87 < x < 0.88$

4. $\tan x = x$

$1.57 \sim \pi/2$ and $4.71 \sim 3\pi/2$

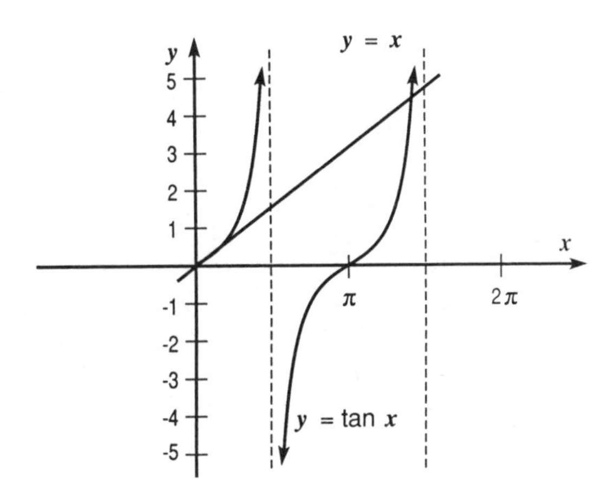

In this region, the graph of $y = x$ crosses the graph of $y = \tan x$ at approximately $x = 4.49$

Let $f(x) = x - \tan x$

Note: Be sure that calculator is in radian mode!

x	$f(x)$
4.2	2.4
4.6	−4.26
4.4	1.3
4.5	−0.14
4.48	+0.25
4.49	+0.068

$4.49 < x < 4.50$

5. $2x = 5 + \sin x$, $2x - 5 = \sin x$

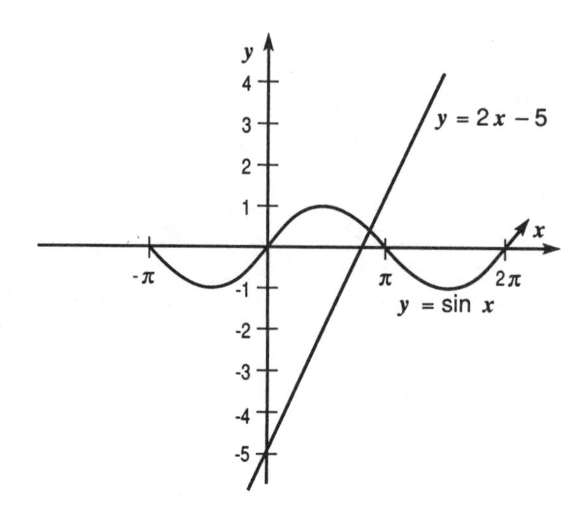

The graph of $y = 2x - 5$ crosses the graph of $y = \sin x$ at approximately $x = 2.7$.

Let $f(x) = \sin x - 2x + 5$

Note: Be sure that calculator is in radian mode!

x	$f(x)$
2.6	−0.32
2.8	+0.26
2.7	−0.03
2.72	+0.03
2.71	+0.0017

$2.70 < x < 2.71$

6. $2x = e^{-x}$

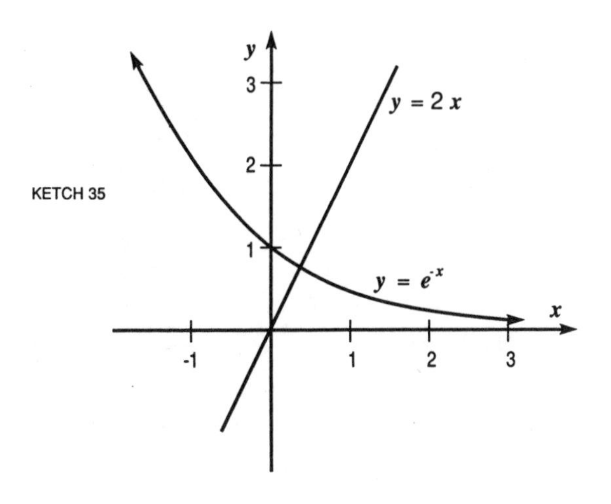

The graph of $y = 2x$ crosses the graph of $y = e^{-x}$ at approximately $x = 0.3$.

Let $f(x) = 2x - e^{-x}$

x	$f(x)$
0.2	-0.42
0.4	$+0.13$
0.3	-0.14
0.35	-0.0046
0.36	$+0.022$

$0.35 < x < 0.36$

7. $2x^3 + e^x = 0,\ 2x^3 = -e^x$

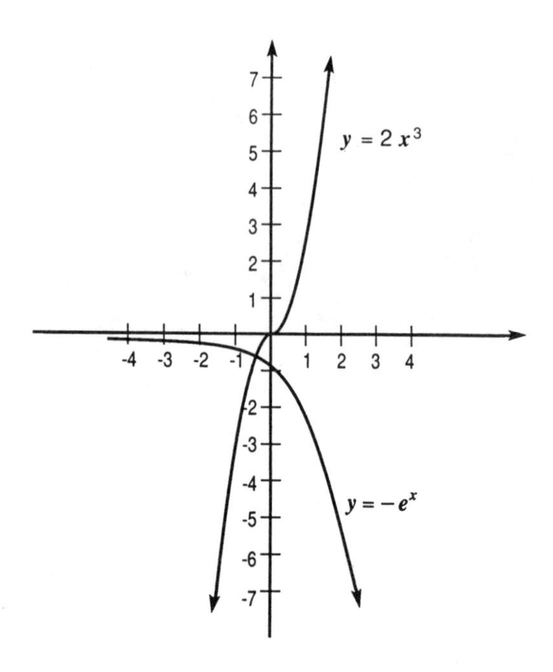

The graph of $y = 2x^3$ crosses the graph of $-e^x$ at approximately $x = -0.6$.

Let $f(x) = 2x^3 + e^x$

x	$f(x)$
-0.7	-0.19
-0.5	$+0.36$
-0.6	$+0.12$
-0.67	-0.090
-0.66	-0.058
-0.64	$+0.003$
-0.65	-0.027

$-0.65 < x < -0.64$

Section C

1. $V = x(50 - 2x)(36 - 2x)$
$= 4x^3 - 172x^2 + 1800x$
$= [(4x - 172)x + 1800]x$

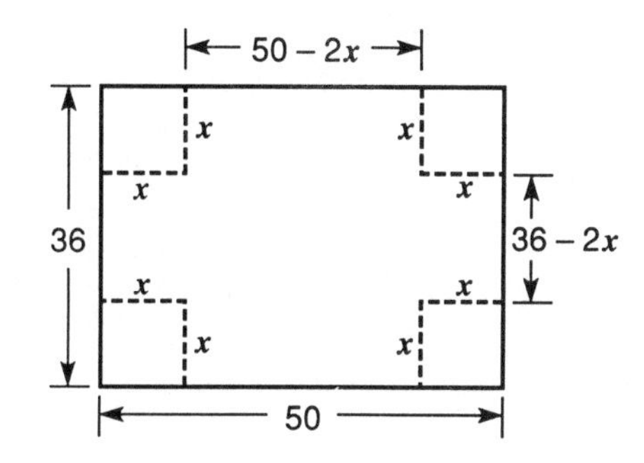

Clearly, $0 < x < 18$

x	V
0	0
3	3960
6	5472
9	5184
12	3744
15	1800
18	0

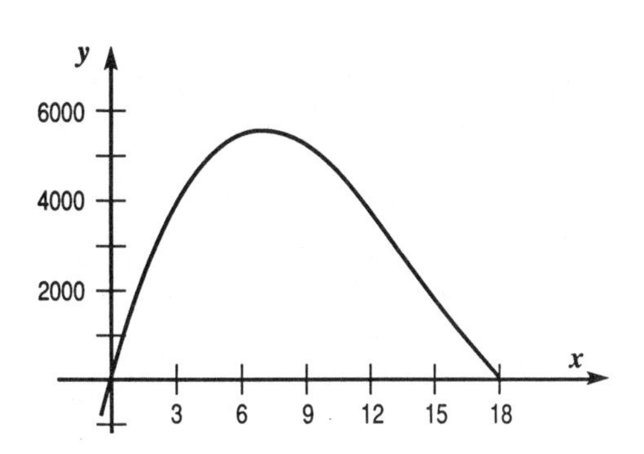

The table above indicates maximum V is obtained for $x \in (3,9)$.

x	V
5	5200
7	5544
8	5440
6.5	5531.5
7.5	5512.5
6.8	5544.45
6.9	5545.12

To the nearest 0.1 cm, $x = 6.9$ cm is the size of the corner that should be cut from the rectangle to obtain the box of largest volume.

Note: The problem above could be worked exactly using methods of calculus.

2. a) and b)

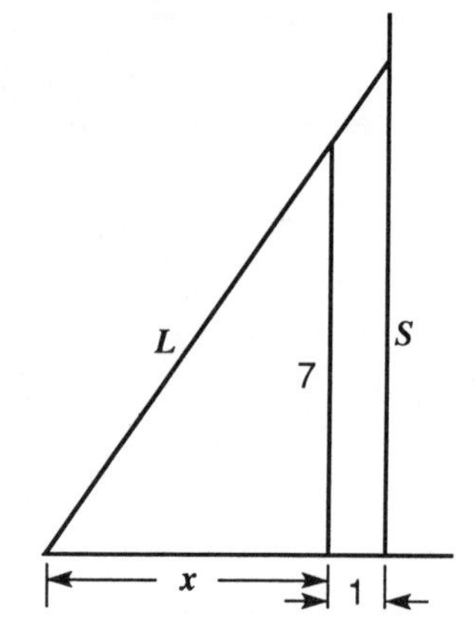

c) $\dfrac{s}{x+1} = \dfrac{7}{x}, \quad s = \dfrac{7(x+1)}{x}$

d)
$$L^2 = (x+1)^2 + s^2$$
$$= (x+1)^2 + \left[\frac{7\,(x+1)}{x}\right]^2$$
$$= \frac{x^2\,(x+1)^2 + 49\,(x+1)^2}{x^2}$$
$$= \frac{(x+1)^2\,(x^2+49)}{x^2}$$

$$L = \sqrt{\frac{(x+1)^2\,(x^2+49)}{x^2}}$$
$$= \frac{x+1}{x}\sqrt{x^2+49}, \text{ since } x > 0$$
$$= (1 + 1/x)\sqrt{x^2+49}$$

AL or *AOS*
value of x [STO] [x^2] [+] 49 [=] [$\sqrt{\ }$] [EXC] [1/x] [+] 1 [=] (unnecessary in AL) [×] [RCL] [=] Display shows value of L

e)

x	1	2	4	6	8
L	14.14	10.92	10.08	10.76	11.96

f) For $x = 4$, the value of L was smallest.

g) The shortest possible ladder is 10.05731 feet (i.e., to the nearest 0.00001 ft!).

h) $2 < x < 6$

i)

x	L
2	10.92
4	10.077822*
6	10.76
3	10.15
5	10.32
3.5	10.062306*
3.7	10.057621*
3.6	10.057986
3.8	10.060955
3.65	10.057326*
3.64	10.05738
3.67	10.057331
3.66	10.057309*

* indicates smallest value of L at that point of computation.

To the nearest 0.01, $x = 3.66$ feet.

3. b) $\frac{14 - h}{r} = \frac{14}{5}$

c) $h = 14 - \frac{14r}{5}$

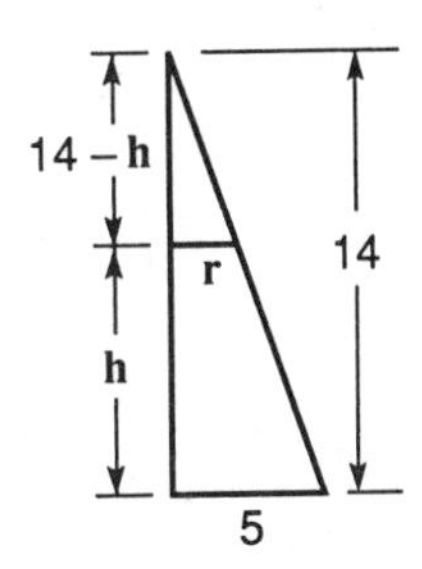

d) $V = \pi r^2 h = \pi r^2 \cdot \left[14 - \frac{14r}{5}\right] = \left[1 - \frac{r}{5}\right] \cdot 14\pi r^2$ (for easier computation)

e) From the physical constraints of the problem, $0 < r < 5$.

r	V	
0	$0(14\pi)$	
5	$0(14\pi)$	
2	$2.4(14\pi)$	
4	$3.2(14\pi)$	
3	$3.6(14\pi)$*	
3.5	$3.675(14\pi)$*	
3.3	$3.7026(14\pi)$*	
3.4	$3.6992(14\pi)$	
3.2	$3.6864(14\pi)$	
3.35	$3.703425(14\pi)$*	
3.33	$3.7036926(14\pi)$*	(= 162.89691)
3.32	$3.7035264(14\pi)$	(= 162.8896)
3.34	$3.7036592(14\pi)$	(= 162.89544)

Note: Since 14π is a constant, we choose to simply indicate the product (14π), thus saving keystrokes. Alternatively, each indicated product may be calculated with the same result. These latter values are shown in parentheses in the last three lines.

* indicates largest value of V at that point of calculation.
To the nearest 0.01, $r = 3.33$ inches gives the largest volume.

f) $h = 14 - \frac{14r}{5} = 14 \cdot \left[1 - \frac{r}{5}\right] = 14 \cdot \left[1 - \frac{3.33}{5}\right] = 4.68$ inches.

Note: Since r has three significant figures, we are justified in leaving only three significant figures in h. To verify this, calculate h for $r = 3.32$ and for $r = 3.34$.

4. b) $12 + x/5$ cents = cost to run truck 1 mile

c) $d = rt$
$1 = xt$
$t = 1/x$

d) Driver earns $6 an hour. In $t = 1/x$ hours the driver earns $6/x$ dollars.

e) Since cost in (b) is in cents and cost in (d) is in dollars, let driver cost $= 600/x$ cents.
[Alternatively, let truck cost $= 1/100(12 + x/5)$.]

f) total cost $= 12 = x/5 + 600/x$

g)

x	total cost (in cents)
40	35
50	34*
60	34*
70	34.57
55	33.909*
54	33.911
56	33.914

The optimal speed at which to run the truck in order to minimize costs is 55 mph.